1982

DNA, Chromatin and Chromosomes

DNA, Chromatin and Chromosomes

E. MORTON BRADBURY

BSc, PhD

Department of Biological Chemistry
University of California, Davis

NORMAN MACLEAN

BSc, PhD

Department of Biology
University of Southampton

HARRY R. MATTHEWS

BSc, PhD

Department of Biological Chemistry
University of California, Davis

A HALSTED PRESS BOOK

JOHN WILEY & SONS INC

NEW YORK – TORONTO

First published 1981

Printed and bound in Great Britain

Published in the U.S.A. and Canada
by Halsted Press,
a Division of John Wiley & Sons, Inc.,
New York

Library of Congress
Cataloging in Publication Data

Bradbury, Edwin Morton.
    DNA, chromatin and chromosomes.

    'A Halsted Press book.'
    Includes index.
    1. Chromatin. 2. Deoxyribonucleic acid.
3. Chromosomes. 4. Gene expression.
I. Maclean, Norman, 1932–
II. Matthews, Harry Roy, 1942–
III. Title. [DNLM: 1. DNA.
2. Chromatin. 3. Chromosomes.
QU 58 B1978d]   QH599.B7   574.87'32
81-3321
ISBN 0–470–27173–6   AACR2

Contents

Preface

This book is chiefly about chromatin, which is the packaged form of the genetic material found in all higher organisms. Knowledge of chromatin, its structure, replication and activity, has increased greatly in recent years, and the book endeavours to provide a reasonably up to date account of present understanding in this area for both undergraduate and postgraduate students. Within a living cell the crucial element of chromatin is of course the DNA. In addition, for certain critical stages in the life of the cell, the chromatin itself is organized into discrete blocks, the chromosomes. It follows that these three aspects of genetic organization are inevitably bound up together, and chromatin cannot be usefully or seriously discussed without repeated reference to the biochemistry of DNA on the one hand, and the biology of chromosomes on the other. Thus this book emerges as one devoted to a unified triad of topics, DNA, chromatin and chromosomes, but with a central emphasis on the middle element.

Our plan in writing the book has been to combine our varied interests in DNA, chromatin and gene expression to form a relatively coherent account which no one of us could have undertaken on our own. The task has been rendered more exciting but less straightforward by the current fast rate of change in the field of molecular genetics, and the ground has more than once shifted under our feet while the book was being written. Nevertheless, it has proved an interesting exercise for us and we hope that some of the stimulation that we have found in the rapid progress of the subject will be carried over to those who read and use the book.

Many of our friends and colleagues have been of great assistance in its preparation, especially Dr. M. Ashburner, Professor M. Callan, Dr. T. Drabble, Dr. S. Gregory, Dr. V. Hilder, Dr. A. MacGillivray and Professor H. Mac-Gregor. We also owe a debt of gratitude to Robert Campbell and the staff at Blackwell Scientific Publications who have patiently collaborated in the production of this book.

Summary of Chapters

CHAPTER 1 COMPONENTS OF EUKARYOTIC CHROMATIN

Eukaryotic chromatin consists mainly of DNA, histones and non-histone proteins (NHP) with small amounts of RNA and other components. DNA occurs as a double stranded helix with complementary bases paired and the base-pairs stacked in the centre of the helix. DNA can adopt a variety of structures with different angles of the base pairs to the helix axis and different pitches to the helix. DNA adopts the B-form in free solution with the bases at right angles to the helix axis and 10.4 base pairs per turn of the helix. It is possible that specific DNA sequences, especially when complexed with specific proteins, might form other structures. Histones are basic proteins which fall into five major classes, H1, H2A, H2B, H3 and H4. They are responsible for maintaining the basic repeating structure of chromatin. Histones H3 and H4 have highly conserved amino acid sequences. The sequences are highly assymmetric with very basic N-terminal regions and aromatic, apolar and acidic residues in the C-terminal regions. Histones H2A and H2B are also highly conserved although they vary more than H3 and H4. H2A and H2B have very basic terminal regions with the apolar residues clustered in the central sections. Histone H1 at about 220 residues is about twice as large as the other histones. Its sequence is also more variable although a central, hydrophobic, region is conserved. The terminal regions are highly basic, especially the C-terminal half of the molecule. Histones form a specific octameric complex containing 2 each of H2A, H2B, H3 and H4 which forms the protein core of the nucleosome. Histones can be modified chemically after synthesis by phosphorylation, acetylation, methylation and diphosphoribosylation at specific sites in the amino acid sequence. Phosphorylation and acetylation are involved in conformational transitions of chromatin. Non-histone proteins are also involved in conformational transitions in chromatin. The HMG non-histone proteins are being characterized and may be involved in the structure of transcriptionally active chromatin.

CHAPTER 2 CHROMATIN STRUCTURE

The basic sub-unit of chromatin structure is called the nucleosome. It contains nine histone molecules namely the octamer core plus one H1, with approximately 195 base pairs of continuous DNA. The DNA length varies between organisms, between cell types and within one cell type. Within the nucleosome the octamer core is closely associated with 145 base pairs of DNA to form the core particle, which does not vary. Non-specific nucleases have been used to isolate nucleosomes and core particles and to probe their structure. Deoxyribonuclease-1 makes single strand breaks ('nicks') preferentially at sites about 10.4 bases apart, implying the DNA is coiled on the outside of the nucleosome in a modified B form structure. Deoxyribonuclease-1 preferentially attacks DNA in chromatin that is being transcribed. Neutron scattering techniques can distinguish between DNA and protein in the core particle and have been used to show that the core particle is a flat disc about 6 nm thick and about 11 nm in diameter. The outer part of the disc is mostly DNA; the inner part protein. X-ray diffraction studies of crystals of core particles lead to a similar model and should provide further details in future. The main features of the core particle structure can be obtained with only H3 + H4 + DNA. Nucleosomes can coil into a filament of 33 nm diameter which is stabilized by Mg^{2+} ions and by histone H1. Further coiling or packing is thought to occur, maybe stimulated by phosphorylation of H1. The final stage of chromosome packing occurs in the metaphase chromosome where loops or domains of DNA are organized on a protein matrix or 'scaffold'. Some features of metaphase chromosomes can be revealed by specific staining or 'banding' patterns. The conformational transitions of chromatin probably involve histone acetylation and phosphorylation.

CHAPTER 3   THE GENOMES OF BACTERIA, VIRUSES, PLASMIDS, MITOCHONDRIA AND CHLOROPLASTS

In comparison with eukaryotes, prokaryotes and subcellular genetic factors have very small genomes which are consequently utilized in a very economic way. *E. coli* probably has fewer than 1800 genes and viruses anything from 3 to 150. Gene regulation in prokaryotes is largely by operon-like arrangements of genes, and DNA synthesis is often continuous in a growing culture, both features distinguishing these cells from eukaryotic counterparts.

The organization of prokaryotic DNA to form a protein/DNA chromatin complex is probable although still somewhat uncertain, but viruses resident in the nuclei of eukaryotic cells are clearly organized into nucleosome complexes by utilizing the normal eukaryotic histones.

Economies practised by bacteria and viruses in utilizing their genomes are interesting and numerous, including regular use of both strands of a double stranded DNA molecule, common leader sequences for some 'mosaic' viral messenger RNAs, and overlapping transcription of the same DNA to yield different proteins by shifting the reading frame. The genomes of some viruses have now been entirely sequenced, as have quite large tracts of the *E. coli* genome.

CHAPTER 4 SEQUENCE AND GENE ARRANGEMENTS IN DNA

The DNA content of a haploid cell (the C-value) varies widely between bacteria and higher eukaryotes. In bacteria it corresponds roughly to the amount of DNA required to code for proteins and immediate control sequences but in higher eukaryotes the C-value is much larger than required to code directly for the 5000 to 50 000 genes expected in a higher eukaryote. DNA reassociation experiments show that bacterial DNA has largely single copy DNA sequences but higher eukaryotes have substantial amounts of reiterated DNA sequences, which fall into three classes: middle repetitive DNA; highly repetitive DNA and inverted repeat DNA. Some of these sequences are clustered and can be isolated as satellites in equilibrium density gradients. Some highly repetitive DNA's have been sequenced and have a short basic repeat of 4 or more bases which is repeated with minor variations. They can often be located in metaphase chromosomes by in-situ hybridization and many such sequences occur at the centromeres or at the ends of the chromosomes. In most higher eukaryotes the middle repetitive DNA is interspersed in short stretches with the single copy DNA although *Drosophila* has a different interspersion pattern with long stretches of uninterrupted single copy DNA. Some middle repetitive DNA is repeated genes but the function of the remainder is unknown. They may be involved in control processes. Nucleic acid sequences can be determined directly by using restriction nucleases to prepare specific fragments and end labelling with base specific degradation or interrupted synthesis to sequence the fragments. Sequences of up to several thousand base pairs have been determined in this way. Study of specific DNA sequences has been greatly facilitated by the ability to clone a specific sequence in a virus or plasmid vector and so prepare relatively large amounts of highly purified material. Detailed study of nucleic acid sequences has revealed some unexpected features, particularly overlapping genes in viruses, interrupted coding sequences in eukaryotes ('split genes') and rearrangement of DNA sequences during development.

CHAPTER 5 THE CELL CYCLE AND REPLICATION

During normal cell proliferation the chromosomes undergo mitosis which involves chromosome condensation, separation and de-condensation. The rest of the cell proliferation cycle is divided into G1, S and G2 phases. Most histone and DNA synthesis occurs in S phase but there is probably a sequence of structural transitions in chromatin throughout the whole cycle. The normal diploid cell, in eukaryotes, can be converted into haploid cells, for reproduction, by meiosis. Crossing over and genetic recombination occur during mitosis. DNA synthesis is carried out by a semi-conservative procedure in which each new strand is 'copied' from one complementary pre-existing strand to give a double strand containing one old and one new single strand. Enzymatically, the process is complex with the DNA being synthesized in short segments which are then joined. The structural changes in chromosomes that occur during the cell cycle and during differentiation are correlated with post-synthetic modifications of histones, particularly phosphorylation of H1 and acetylation of H3 and H4. Phosphorylation of H1 at specific sites probably controls the packing of coils of nucleosomes, particularly during the initial stages of mitosis. Acetylation of H3 and H4 probably affects nucleosome interactions to allow transcription to occur. Transcriptionally active chromatin displays a nuclease resistant DNA repeat similar to that in bulk chromatin. However, the nucleosome structure of transcriptionally active chromatin is different, particularly in its high sensitivity to DNase-I and rapid degradation to monomer nucleosomes by micrococcal nuclease.

CHAPTER 6 TRANSCRIPTIONALLY ACTIVE CHROMOSOMES

Both the lampbrush chromosomes of vertebrate oocytes and the giant polytene chromosomes of some Dipteran larval tissues are transcriptionally active, and have therefore made an outstanding contribution to our knowledge of gene function. Study of these structures has revealed that, in the first case, the active structural genes are located on the loops of DNA which are attached to the chromosome axis, and in the second, that they are found within the banded regions of the chromosomes. The correlation between genes and the loops or bands of these chromosomes is strong, but not absolute. The fact that these structures number some 2000 to 8000 per genome suggests that the true number of structural genes in these organisms may be close to these approximates. At a functional level, the two chromosome types suggest different models, since in the lampbrush chromosome most or all of the loops are active transcriptionally, but in the polytene chromosome puffing is restricted to a few bands at any one time. It is probable that the latter represents the more typical situation.

In the next chapter these models of chromosome function are discussed along with other information in a general consideration of gene regulatory mechanisms.

CHAPTER 7 CHROMATIN ACTIVITY AND GENE REGULATION

In this chapter we attempt to draw together the scattered and obviously partial evidence about gene regulation, together with the main theoretical contributions to this topic. While the operon model serves to illustrate effectively the normal mechanisms of gene regulation in prokaryotes, it is now obvious that it is not generally applicable to the eukaryotic situation. In higher cells quite a lot of the emphasis in gene control is at a post-transcriptional level i.e. processing of RNA. The transcriptional level of control is presumed to be mediated by special gene regulatory molecules, RNA or protein, which coordinate and specify the precise aspects of gene expression which constitute the basic mechanisms underlying eukaryotic cell differentiation.

Abbreviations

ATP	adenosine triphosphate
Å	angstrom unit
bp	base pair
cDNA	complementary DNA
Hn RNA	heterogeneous nuclear RNA
HMG	high mobility group
kb (1000 bases length of DNA)	kilobases
Mdal	megadaltons (1 megadalton $= 10^6$ daltons)
mRNA	messenger RNA
mol wt	molecular weight
10^{-9} gram (ng)	nanogram
10^{-9} metre (nm)	nanometre
10^{-12} gram (pg)	picogram
pfu	plaque-forming unit

Scale for conversion between kilobase pairs or duplex DNA and molecular weight.

Kilobase pairs

Megadaltons

Chapter 1
Components of Eukaryotic Chromatin

1.1 INTRODUCTION

So far most of our detailed understanding of how genes function has come from studies of prokaryotes. These are free-living unicellular organisms (bacteria and blue-green algae), in which the genetic material is distributed throughout the cell organism. Both bacteria and viruses are well-suited for laboratory investigation and the powerful combination of biochemistry and genetics has led to most of our current concepts concerning the primary structure, control and expression of their genes. Prokaryotes are the most simple forms of living organisms. Higher life forms, called collectively, eukaryotes, can be distinguished from prokaryotes by several features, in particular by the presence of a cell nucleus which separates their chromosomes from the cytoplasm. Also whereas we can roughly equate the DNA content of prokaryotes with the number of genes they are known to contain this is not the case with eukaryotes. Although there is a very crude correlation between DNA contents and the genetic complexities of eukaryotes it is a general observation that eukaryotes contain appreciably more DNA than can be accounted for by the number of genes thought necessary to specify the organism. The functions of the very large proportions of non-coding DNA are not understood at present, though as will be seen later there have been unexpected and very exciting discoveries concerning the complexity of eukaryotic DNA. A third major difference between prokaryotes and eukaryotes is the presence in the latter of a group of basic proteins called the histones which are strongly complexed with DNA in chromosomes.

There is an arbitrary division of eukaryotes into lower and higher eukaryotes. Nucleated single cell organisms such as yeasts, algae, and protozoa, are regarded as lower eukaryotes, as are simple forms of multicellular organisms, for example moulds and other simple fungi. Higher eukaryotes usually refer to what are recognizable as animals and plants. Between these two extremes there is a multitude of organisms and it would not be useful to classify them other than as eukaryotes.

Probably one of the most important series of unsolved problems in biology is concerned with the organization and function of eukaryotic chromosomes. There are closely interrelated structure–function problems: the sequence organization of eukaryotic DNA; the structure of inactive chromatin; the structural transition which occurs when genes become active and are transcribed; the determination, maintenance and control of active genes in differentiated tissue; the control of chromosome structure through the cell cycle; the dis-

1

assembly and reassembly of chromatin during DNA replication; the structure
of metaphase chromosomes, etc.

Recently there have been major advances in this area of biological research
and the solutions to several of these problems are now in sight. This fortunate
situation is a consequence of major advances in biochemical techniques and
preparative strategies.

1.2   COMPONENTS OF EUKARYOTIC CHROMATIN

Historically chromatin was the term used to describe the contents of interphase
nuclei as visualized in the light microscope. More recently chromatin has been
adopted by biochemists as a label for the deoxyribonucleoprotein complex
isolated from cells and tissues.

The composition of eukaryotic chromatin is illustrated in Fig. 1.1. There
are four major components: DNA; a group of five classes of basic proteins,
histones, which are now regarded as major structural proteins of eukaryotic
chromosomes; a large number of other chromosomal proteins so far poorly
characterized and called non-histone proteins (NHP); and RNA. Some general
rules have emerged from studies of the compositions of the nuclei of different
cells. Firstly, the ratio of the amount of histone to that of DNA is relatively
constant for all tissues and lies in the range from 1.1 to 1.3. This constancy and
the close association of histones with DNA has led to many proposals concerning
their role in all aspects of chromosome structure and function. Secondly,
although the amounts of NHP are very variable with NHP:DNA ratios in the
range 0.2–0.8 there appears to be a rough correlation of these amounts of NHP

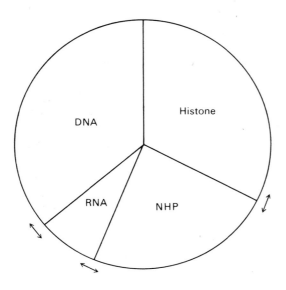

Fig. 1.1 Components of the
eukaryotic chromatin. The
histone to DNA ratio is kept
constant at about 1.1 to 1.3.
The arrows indicate variability
in the amounts of RNA and
NHP depending on the
metabolic activity of the cell.

with the metabolic activity cells: the higher the metabolic activity the larger the proportion of NHP. A similar positive correlation appears also to hold for the variable low content of RNA.

It is not clear at present which of the NHP have roles in chromosome structure and function. Chromatin has the ability to bind non-chromosomal proteins during the isolation procedures. There is evidence to suggest, however, that the proteins responsible for control of genetic activity reside in the NHP component though these important regulatory proteins will be difficult to identify because they are expected to be present in very small amounts. Other components of the NHP include the enzymes required for the processing of DNA and for other nuclear functions; proteins required for the structural differences between inactive and active chromatin and those required for the integrity of metaphase chromosomes.

1.3 COMPLEXITY OF EUKARYOTIC DNA

A major enigma in the molecular biology of higher organisms is posed by the very large amounts of DNA found in the eukaryotic genome. The amount of DNA in the nucleus of the human sperm cell is 3.2 pg which corresponds to a total linear length of the DNA molecules of 1.9 m. This is not the largest amount of DNA found in the cells of higher organisms, e.g. the nucleus of the plant *Tradescantia* contains approximately 18 m of DNA. If it is assumed that all of the DNA in the nucleus of the human cell is coding for proteins then taking the average size of a protein as 300 amino acids it can be estimated that 1.9 m of DNA would be sufficient to code for 6×10^6 genes. From genetic arguments, however, it is thought that 50 000 genes are sufficient to specify man and these would require about 5% of the total complement of human DNA. It is now generally accepted that only a small fraction of DNA in eukaryotes actually codes for proteins and possibly as much as 90% of the DNA is involved in other functions, e.g. for the control of the expression of genes in differentiated tissues as well as possible structural roles in the organization of eukaryotic genomes. The analysis of eukaryotic DNA will be discussed in more detail in Chapter 4 where it will be shown that the DNA can be classified into different types, some of which can be correlated with different function. In addition to this genetic complexity these large amounts of eukaryotic DNA have to be controlled during such processes as transcription, DNA replication and packaging into the metaphase chromosome. Mechanisms must exist, therefore, to control these enormous lengths of DNA in nuclei and also for the precise and reversible packaging of DNA into the metaphase chromosome. We are now in a position where we can discuss some of the mechanisms proposed for these processes. Firstly, we shall consider some of the properties of DNA, histones and NHP relevant to these proposals and then discuss in detail current ideas on the molecular structure of chromatin.

1.4   STRUCTURE OF DNA

DNA is the best known of all biological structures both to the layman and biology student because of its central position in biology as the library of genetic information and the blueprint for the development of organisms. The base pairing scheme adenine–thymine (A–T) and guanine–cytosine (G–C), now widely illustrated in school textbooks, led to the solution of the genetic code whereby information contained in the DNA base sequence is first transcribed into messenger RNA and then translated to protein sequences. In fact the feature of the Watson–Crick DNA structure which had the greatest impact on biology in 1953 was the base pairing scheme rather than the attractive symmetrical double stranded helical structure. Now more than 25 years later it is the physical structure of DNA which has assumed major importance as we consider the mechanism by which DNA is controlled in the nucleus.

Historically, there are three well-defined structures of DNA called the A-, B- and C-forms. These forms are found in fibres of pure DNA and depend on the nature of the cations and the degree of hydration of the DNA. Their structures were determined from X-ray diffraction studies of fibres of various salts of DNA, e.g. sodium DNA, and lithium DNA at different water contents. It should be pointed out however that fibres are not single crystals and as a result there are limitations to fibre diffraction studies of DNA. The molecules in a single crystal are precisely arranged in three dimensions and X-ray diffraction studies can lead to a completely unambiguous determination of the structure of the molecule at atomic resolution. In comparison, when a fibre is drawn from DNA or any long extended biological macromolecule the fibrils or microcrystals are not perfectly aligned and the axes of the fibrils are contained within a cone of angles about the fibre axis. For a highly ordered fibre most of the DNA molecules would be contained within a very small cone of angles although they have all possible orientations around the fibre axis. Thus, whereas the structural information in a crystal can be separated along three axes, the information in an X-ray fibre diffraction pattern can be separated in only two dimensions; along the fibre axis and at right angles to the fibre axis. It follows that in general a fibre structure determination is not unambiguous. The correctness of a model obtained from fibre diffraction studies depends on the crystallinity and on the degree of orientation of the molecules in the fibre which gives the total sum of information contained in the fibre diffraction pattern. These general points can be illustrated by the X-ray diffraction patterns shown in Fig. 1.2. The single crystal pattern of tyrosyl tRNA synthetase contains many hundreds of separated sharp diffraction spots and this is only a small part of the total information. In contrast the fibre diffraction pattern of the A-form of DNA contains far fewer more diffuse diffraction spots and many of these spots are the sum of overlapping or coincident reflections. This is the total diffraction data available to determine the structure of the A-form DNA. The processes by which fibre structures are

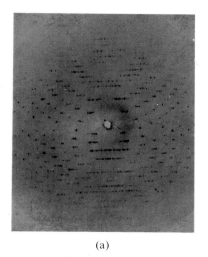
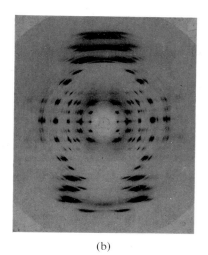

(a)   (b)

Fig. 1.2 Comparison of (a) a low angle 2.5° oscillation X-ray diffraction pattern of tyrosyl tRNA synthetase (Provided by Prof. D. M. Blow, F.R.S. and published in Reid B. R. *et al.* (1973) *J. Mol. Biol.* **80**, 199) with (b) the fibre X-ray diffraction patterns of the A-form DNA. (Courtesy of Prof. Strather Arnott.)

determined involve trial and error. However, because we are dealing with a very long stiff molecule and there is available a substantial amount of stereo-chemical information from single crystal studies of nucleosides and nucleotides, the number of possible conformations of the molecule is considerably reduced. The best conformation is determined by systematically calculating the diffraction pattern of the model structure and comparing this with the observed diffraction pattern. The conformation which gives the best fit between observed and calculated fibre diffraction patterns is the 'accepted' model for the structure. It is however 'accepted' only so long as no better model is found to fit the experimental data.

Arnott and colleagues have made systematic studies of the various conformations of DNA and synthetic polynucleotides. At least seven conformations have been characterized and they fall into two classes of DNA conformations called A and B after the classic A and B structures of DNA. The C-form of DNA is now known to be a variant of the B-form and thus belongs to the B-class of conformations. The major difference between the two classes is the pucker of the furanose sugar ring. Models for the A-, B- and C-forms are given in Fig. 1.3. The similarity of the B- and C-forms is clear.

Projections along and down the helix axis of the B-form DNA show the well-known features of this molecule. The deep (or wide) and shallow (or narrow) grooves around the double helix are clear. The two polynucleotide chains have opposite polarity, i.e. they point in opposite directions. This can best be seen in the longitudinal projection from the orientations of the sugar

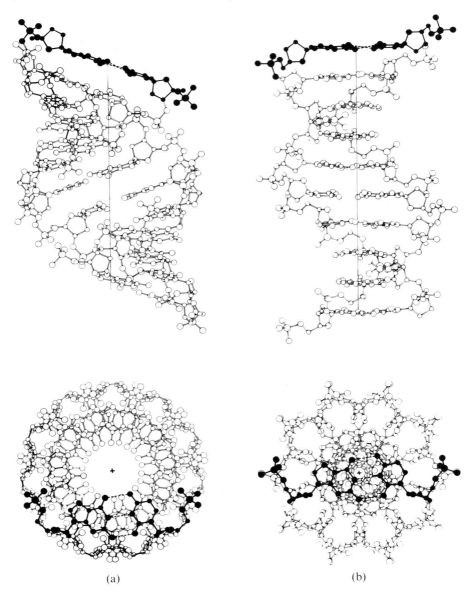

(a)   (b)

rings in the top nucleotide pair which define the 5′ and 3′ ends of the polynucleotide chains. The sugar ring 'points' up at the 5′ end of the polynucleotide
chain and down at the 3′ end of the complementary polynucleotide chain. The
planes of the hydrogen bonded base pairs are perpendicular to the helix axis
and adjacent base pairs are separated by 0.34 mm. In both projections the conformation of a G–C nucleotide pair is emphasized. The secret to solving the

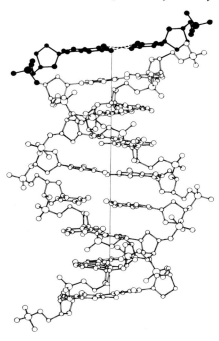

Fig. 1.3 Models for the (a) A form, (b) B form and (c) C form of DNA.

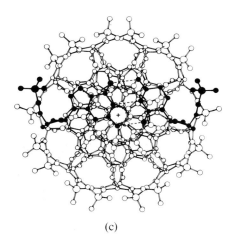

(c)

structure of DNA was the finding that a G–C base pair was equivalent to an A–T base pair in the helix structure.

Because there are two polynucleotide chains in DNA and they have opposite polarity it is necessary to define the linear representation of a DNA base sequence. A single stranded DNA sequence is written for the non-coding DNA strand from the 5' to the 3' end. For double stranded DNA the upper line of base

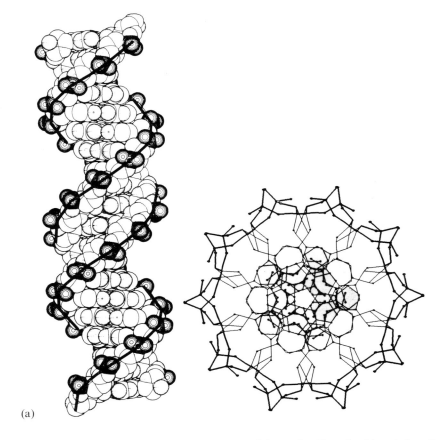

(a)

sequence also refers to the non-coding strand from the 5' to the 3' end. In effect the DNA sequence written in this way defines the transcribed RNA sequence. It follows that the DNA coding strand is transcribed from the 3' to the 5' end.

The A, B and C double helical conformations are characterized by two parameters; T, the rotation about, and H, the translation distance along the helix axis between successive nucleotides. These parameters are accurately determined from the fibre diffraction patterns. For the classic A-form $T = 32.7°$ or 11.0 base pairs per turn of the helix and $H = 0.256$ nm to give a pitch of the helix of 2.82 nm; for the classic B-form $T = 36.0°$ or 10.0 base pairs per turn of the helix and $H = 0.34$ nm to give a pitch of the helix of 3.4 nm. For the classic C-form $T = 40.0°$ and $H = 0.328$ nm i.e. 9.0 base pairs per turn and a helix pitch of 2.95 nm. Another feature often referred to when describing the different conformations of DNA is the angle the planes of the bases make with the helix axis, as shown in Fig. 1.3 these angles are found to be close to 90° for the B-form, 85° for the C-form and 70° for the A-form.

Some general rules have emerged from these systematic studies concerning

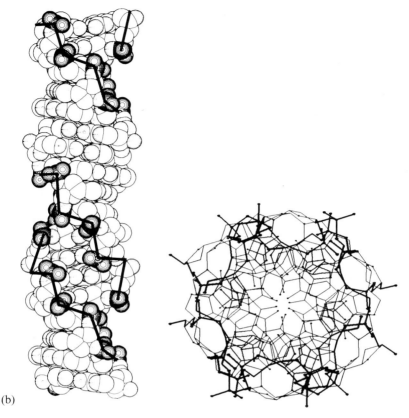

(b)

Fig. 1.4 Space-filling drawings of (a) the B- and (b) Z-forms of DNA and end-views of the same. (In Wang A. H. *et al.* (1979) *Nature* **282**, 680.) Kindly provided by Professor Alexander Rich.

the factors determining the different conformations of DNA. DNA is in the B-form at high hydration in fibres. In aqueous solution DNA will be in the B-type structure, close to but not necessarily identical with the fibre B-form DNA. The reason for caution is that in fibres DNA molecules are constrained by packing requirements and the precise helical form of DNA may be determined by intermolecular interaction within the crystallite. These constraints are removed in aqueous solution and the minimum energy conformation of the DNA may be slightly different from the fibre form; X-ray scattering data on DNA in solution suggest that there are almost 10.7 base pairs per turn. This is now thought to be high and current estimates indicate that the B-form DNA in solution has between 10.4 and 10.5 base pairs per turn of the helix. Such considerations are important when considering the structure of the DNA in chromatin.

When the state of hydration of DNA is reduced e.g. by decreasing the water

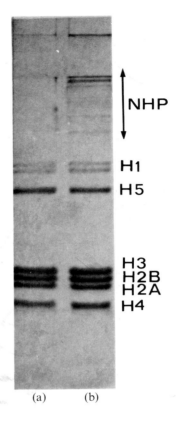

NHP

H1
H5

H3
H2B
H2A
H4

Fig. 1.5 Polyacrylamide gel patterns of histones from (a) chicken erythrocyte chromatin and (b) chicken erythrocyte nuclei.

(a)   (b)

content of fibres, the conformation of DNA can undergo transitions from the B-form to the C-form and then to the A-form or, if the DNA has very high contents of A–T or G–C, to the D-form. The D-form belongs to the B-class of conformation and has $T = 45°$ and $H = 0.303$ nm i.e. 8.0 base pairs per turn and a pitch of 2.42 nm. Such transitions are of interest when considering the problems of protein recognition of DNA sequences. It must be remembered that in the cell nucleus DNA is in association with proteins. Such associations may have the effect of reducing the hydration of DNA so that, depending on the DNA sequence, a variety of localized conformations may be adopted to aid recognition by specific DNA binding proteins.

1.4.1 A New DNA Structure

Until very recently it was thought that any new DNA structure would be a variant of the right-handed double helical A- or B-classes of structure, (Fig. 1.3). It has now been shown unambiguously, that an oligonucleotide with an alternating base sequence d(CpGpCpGpCpGp) i.e. d(CG)$_3$ can take up a distinctly different helical form. In single crystals this oligonucleotide forms a left-handed double helix with Watson–Crick base pairs and antiparallel sugar phosphate chains. Whereas the dC residues have a conformation similar to that found in B-form DNA the dG residues have a quite different conformation with the sugar ring in the C_3 ends form. Thus the dinucleotide CpGp is a structural repeat as well as a chemical repeat. As a result of this alternating conformation of the C and G nucleotides the sugar phosphate chain follows a zig-zag path, hence the name Z-DNA. Space filling models for the B- and Z-forms of DNA are shown in Fig. 1.4. These studies have now been extended to high molecular weight polynucleotides poly d(GC) poly d(GC) and a fibre diffraction pattern corresponding to Z-DNA has been identified for fibres at 44% relative humidity. At higher hydration this alternating polynucleotide also takes up the A and B structural forms. Similar Z-form structures have also been observed with poly d(AC) poly d(GT) suggesting that under certain conditions any DNA segment with an alternating purine–pyrimidine base sequence can assume the Z-helix. Although not yet demonstrated to occur in nature it is possible that this new helical form has a biological function and that segments of alternating purines and pyrimidines may be involved in protein recognition. Further, we can expect with systematic crystallographic studies of other oligonucleotides of specific base sequences that other DNA conformations may be found. Finally it should be stressed that when DNA is strongly complexed with histones in chromatin its structure may also be different from the accepted linear forms of DNA and this will be discussed later.

1.5 CHROMOSOMAL PROTEINS

1.5.1 Histones

Histones are found complexed with DNA in the cells of all higher eukaryotes, except for some sperm cells, and in well-defined lower eukaryotes for which the chromosomal proteins have been biochemically characterized. There have been reports of the absence of certain histones in lower eukaryotes such as tetrahymena, the yeasts, etc. Such reports however are usually associated with organisms from which it is biochemically difficult to extract the chromosomal proteins. When these difficulties have been resolved the full complement of histones is usually found. In most somatic tissues histones form a group of five classes of proteins; H1, H2A, H2B, H3 and H4. In inactive nucleated erythrocytes from birds, fish and amphibians another type of very lysine-rich histone, called H5, has been found in addition to histone H1. H5 is thought either to partially replace or supplement H1 for the purpose of total gene inactivation in nucleated erythrocytes. All these histones can be visualized as bands on a polyacrylamide gel, Fig. 1.5. The amino acid compositions of histones are given in Table 1.1. These show that the histones are highly basic molecules with high contents of lysine and arginine; another feature of histone compositions is

Table 1.1 Amino acid analyses (moles %) of calf thymus histone fractions and chicken erythrocyte H5. (Table provided by Dr F. W. Johns.)

Amino acid	H2A	H2B	H3	H4	H1	H5
Asp (A)	6.6	5.0	4.2	5.2	2.5	1.7
Thr	3.9	6.4	6.8	6.3	5.6	3.2
Ser	3.4	10.4	3.6	2.2	5.6	11.9
Glu (A)	9.8	8.7	11.5	6.9	3.7	4.3
Pro	4.1	4.9	4.6	1.5	9.2	4.7
Gly	10.8	5.9	5.4	14.9	7.2	5.3
Ala	12.9	10.8	13.3	7.7	24.3	16.3
Val	6.3	7.5	4.4	8.2	5.4	4.2
$\frac{1}{2}$-Cys	0.0	0.0	1.0	0.0	0.0	0.0
Met	0.0	1.5	1.1	1.0	0.0	0.4
Ile	3.9	5.1	5.3	5.7	1.5	3.2
Leu	12.4	4.9	9.1	8.2	4.5	4.7
Tyr	2.2	4.0	2.2	3.8	0.9	1.2
Phe	0.9	1.6	3.1	2.1	0.9	0.6
Lys ⎫	10.2	14.1	10.0	11.4	26.8	23.6
His ⎬(B)	3.1	2.3	1.7	2.2	0.0	1.9
Arg ⎭	9.4	6.9	13.0	12.8	1.8	12.4
Acidics (A)	16.4	13.7	15.7	12.1	6.2	6.0
Basics (B)	22.7	23.3	24.7	26.4	28.6	37.9
B/A	1.4	1.7	1.6	2.2	4.6	6.3
Lys/Arg	1.1	2.0	0.8	0.9	15.0	0.8
Gly X Arg	102.0	41.0	70.0	190.0	13.0	65.7
N-terminal group	Acetyl	Proline	Alanine	Acetyl	Acetyl	Acetyl

the absence of tryptophan. Histones have been described according to their lysine or arginine contents. Thus H1 and H5 have been called 'very lysine-rich histones', H2A and H2B 'moderately lysine-rich histones' and H3 and H4 'arginine-rich histones'. It has been known for some years from histone stoichiometries and histone to DNA ratios that there were approximately one of each histone H2A, H2B, H3 and H4 to 100 base pairs of DNA and the stoichiometry of H1 was about half that of the other classes of histones.

1.5.2 Properties of Histones

Analysis of the sequences of the five histones from a variety of tissues and organisms demonstrates two major properties. Firstly, histone sequences, particularly those of histones H3 and H4, are the most rigidly conserved of all proteins so far studied. The remarkable resistance of the histone sequences to change through the evolution of eukaryotes implies that each and every residue is essential for the biological functions of histones. Secondly, all histone sequences exhibit marked asymmetries in the distributions of charged and apolar residues.

To understand these features of histone sequences the properties of amino acids and the different conformations of polypeptide chains should be discussed briefly. The 20 amino acids used as the building blocks of proteins can be divided into groups according to their chemical properties. Apolar residues are so called because they have hydrocarbon sidechains which are hydrophobic in character; examples of these are valine which has the sidechain $[\alpha CHCH(CH_3)_2]$, leucine $[\alpha CHCH_2CH(CH_3)_2]$, isoleucine $[\alpha CHCHCH_3CH_2CH_3]$ and phenylalanine $[\alpha CHCH_2C_6H_5]$. These hydrocarbon sidechains prefer a non-aqueous environment and for this reason apolar residues are usually found in the interior of globular proteins. The stability of globular proteins is due in large part to hydrophobic interactions between these interior apolar residues. Histones are basic because of a high proportion of the positively charged residues lysine $[\alpha CH(CH_2)_4NH_3^+]$ and arginine $[\alpha CH(CH_2)_3NHC(NH_2)_2^+]$. Major interactions between histones and DNA are through the binding of these positively charged sidechains to the negatively charged phosphate groups on the outside of the DNA double helix. Negatively charged residues found in proteins are glutamic acid $[\alpha CH(CH_2)_2COO^-]$ and aspartic acid $[\alpha CHCH_2COO^-]$. Many non-histone chromosomal proteins have a high proportion of acidic residues and as with DNA, could also bind the positively charged histones. In a fourth group of amino acids the sidechains contain aromatic rings; these are tyrosene, tryptophan and phenylalanine mentioned above. Aromatic rings have the potential to interact with the bases in DNA or to intercolate between adjacent base pairs. Of the remaining amino acids serine $[\alpha CHCH_2OH]$ and threonine $[\alpha CHCHOHCH_3]$ form a pair of residues with similar properties and asparagine $[\alpha CHCH_2CONH_2]$ and glutamine $[\alpha CH(CH_2)_2CONH_2]$ form another pair of similar residues. Another way of grouping amino acids is according to their conformational behaviours rather than their chemical properties discussed above. From the detailed three-dimensional or tertiary structures of globular proteins, statistical analyses have been made of the types of secondary structures assumed by the different amino acids. Three major types of secondary structures are found in the structures of globular proteins, these are: the α helix; extended β-chains; random coil conformations. The α-helix is also found in some fibrous proteins, such as α-keratin, and is a regular helix of pitch 0.54 nm with 3.6 residues in one turn of the helix. Extended β-chains are, as the name implies, the almost fully extended structure of the polypeptide chain which is hydrogen bonded to adjacent extended polypeptide chains. The most prevalent form of the β-structure is with adjacent chains antiparallel to each other to give the antiparallel β-sheet. Most insect silks have this extended structure. In a few cases the extended polypeptide chains are parallel to each other and this structure is called the parallel β-sheet. The portion of the polypeptide chain in globular proteins which cannot be ascribed to regular secondary structures is called the random coil. It should be noted, however, that in globular proteins the 'random coil' polypeptide chain is fixed and follows a precise path. The term 'random coil' is also applied to the denatured form of a protein in solution. This is a dynamic flexible random coil form devoid of any regular conformation.

By tabulating the amino acids found in the α-helix, extended B-chain and random coil conformation, it was found that certain residues are found in particular conformations. For example the residues proline, glycine, serine and aspartic acid are called helix-destabilizing residues because proline cannot be incorporated into an α-helix and the other residues are rarely found in this conformation. These helix-destabilizing residues are also found at well-defined reversals of polypeptide chain directions in protein structures called β-bends which also include other residues with small sidechains such as alanine and threonine. β-bends cannot be made with residues with bulky sidechains. The same is also found for the anti-parallel β-structure forms of insect silks which contain very largely residues with the short side chains mentioned above. The preponderance of residues with small sidechains enable β-sheets, made up of hydrogen bonded antiparallel β-chains, to pack in a regular manner to give the required physical properties of insect silks of strength and flexibility. In globular protein this extensive regular packing of β-chains is not found and their β-chains also incorporate residues with bulky sidechains. These β-chains, however, follow the general rule of hydrophobic residues such as valine and phenylalanine towards the interior of the protein and hydrophillic residues, such as aspartic acid, towards the outside in contact with the aqueous medium. Similarly with α-helical segments in globular proteins, the side of the α-helix towards the interior of the protein is hydrophobic and incorporates residues such as leucine, isoleuine, methionine and phenylalanine while the side of the α-helix in contact with the aqueous medium bond includes the hydrophillic residues, lysines, glutamic acids, etc.

Although some of the properties of amino acids described above can be applied to histone sequences it must be kept in mind that nucleic acid binding proteins would be expected to behave differently to 'free living' globular proteins in solution.

1.5.3   Histones H3 and H4

The sequences of histones H3 and H4 are the most highly conserved of all proteins so far studied i.e. these histone sequences have been essentially unchanged through evolution of eukaryotes. This is shown from the sequences of histones H4 from calf thymus and other organisms in Fig. 1.6. There are only two very conservative replacements between bovine (*Bos taurus*) and pea (*Pisum sativum*); valine 60 and lysine 77 in bovine H4 are replaced by isoleucine and arginine respectively in pea. In a conservative replacement a residue is replaced by another residue with very similar properties; both valine and isoleucine are apolar residues with similar hydrocarbon side chains and lysine and arginine are both basic residues. Thus although the lines of evolution of peas and cows probably diverged 1.8×10^9 years ago their H4 sequences are

AcSer — Gly — Arg — Gly — Lys — Gly — Gly — Lys — Gly — Leu[10] —

Gly — Lys — Gly — Gly — Ala — Lys — Arg — His — Arg — Lys[20] —

Val — Leu — Arg — Asp — Asn — Ile — Gln — Gly — Ile — Thr[30] —

Lys — Pro — Ala — Ile — Arg — Arg — Leu — Ala — Arg — Arg[40] —

Gly — Gly — Val — Lys — Arg — Ile — Ser — Gly — Leu — Ile[50] —

Tyr — Glu — Glu — Thr — Arg — Gly — Val — Leu — Lys — Val*[60] —

Phe — Leu — Glu — Asn — Val — Ile — Arg — Asp — Ala — Val[70] —

Thr — Tyr — Thr† — Glu — His — Ala — Lys‡ — Arg — Lys — Thr[80] —

Val — Thr — Ala — Met — Asp — Val — Val — Tyr — Ala — Leu[90] —

Lys — Arg — Gln — Gly — Arg — Thr — Leu — Tyr — Gly — Phe[100] —

Gly — Gly

*Ile in Pea   †Lys in sea urchin, starfish   ‡Arg in Pea.

Fig. 1.6   Sequence of calf thymus histone H4 with variations from other species indicated. (Taken from Von Holt C. *et al.* (1979) *F.E.B.S. Lett.* **100**, 201.)

virtually identical. A similar sequence conservation has also been found for histone H3; compared to bovine H3 the pea H3 sequence has four substitutions, three of them being conservative. The sequence of H3 and H4 exhibit considerable asymmetry in the distribution of different types of residues; the *N*-terminal regions of both H3 and H4 (Fig. 1.6) are very basic containing high proportions of lysines and arginines and a high density of the helix destabilizing residue, glycine. The central and C-terminal regions contain the most of the apolar, aromatic and acidic residues. The protein sequence-conformation correlations described earlier have been applied to the sequences of histones and for histones H3 and H4 they predict that the apolar central and C-terminal regions have a very high potential for secondary structures (i.e. α-helix formation) but not the basic N-terminal regions. This behaviour contrasts with the sequences of globular proteins which do not show such asymmetries.

1.5.4   Histones H2A and H2B

These histones exhibit more sequence variability than H3 and H4. The calf thymus H2A sequence is given in Fig. 1.7. Comparing calf thymus H2A with the trout H2A sequence shows that there are 4 conservative substitutions and 2 deletions; for sea urchin gonad H2A there are 11 substitutions and 4 deletions. All of these sequence differences and the differences between pea and calf H2A

Ac — Ser — Gly — Arg — Gly — Lys — Gln — Gly — Gly — Lys — Ala[10] —

Arg — Ala — Lys — Ala — Lys — Thr — Arg — Ser — Ser — Arg[20] —

Ala — Gly — Leu — Gln — Phe — Pro — Val — Gly — Arg — al[30] —

His — Arg — Leu — Leu — Arg — Lys — Gly — Asn — Tyr — Ala[40] —

Glu — Arg — Val — Gly — Ala — Gly — Ala — Pro — Val — Tyr[50] —

Leu — Ala — Ala — Val — Leu — Glu — Tyr — Leu — Thr — Ala[60] —

Glu — Ile — Leu — Glu — Leu — Ala — Gly — Asn — Ala — Ala[70] —

Arg — Asp — Asn — Lys — Lys — Thr — Arg — Ile — Ile — Pro[80] —

Arg — His — Leu — Gln — Leu — Ala — Ile — Arg — Asn — Asp[90] —

Glu — Glu — Leu — Asn — Lys — Leu — Leu — Gly — Lys — Val[100] —

Thr — Ile — Ala — Gln — Gly — Gly — Val — Leu — Pro — Asn[110] —

Ile — Gln — Ala — Val — Leu — Leu — Pro — Lys — Lys — Thr[120] —

Glu — Ser — His — His — Lys — Ala — Lys — Glu — Lys

Fig. 1.7 Sequence of cal thymus histone H2A. (From Yeoman L. C. *et al.* (1972) *J. Biol. Chem.* **247**, 6018.)

are located almost entirely in the N- and C-terminal regions of the H2A molecules.

The calf thymus H2B sequence has similar properties to H2A. The trout H2B sequence differs from calf thymus H2B by 7 substitutions and 2 deletions; 8 of these changes are contained in the N-terminal 37 residues of the molecule. Similar comparisons have been made of the sequences of histone H2B subfractions from sea urchin sperm and H2B from the mollusc *Patella granatina* and other H2B molecules sequenced by Von Holt and colleagues: in all cases the N-terminal regions contain virtually all of the substitutions, deletions and insertions.

Histones H2A and H2B both have asymmetrical sequences with very basic N-terminal regions and basic C-terminal tails which contain very few apolar residues and virtually all of the sequence differences for these molecules from different sources are located in these basic regions. The central regions of both H2A and H2B contain a high proportion of apolar residues and are very highly conserved, presumably to preserve interactions essential to the functions of these proteins.

1.5.5 Very Lysine-Rich Histones H1 and H5

Histone H1 differs substantially from the other four main histones described above. It is a much larger protein containing between 200 and 250 residues depending on the tissue and organism. H1 is the most variable of the histones, exhibiting heterogeneity within one cell type and substantial variation between cell types and organisms. Rabbit thymus H1 consists of five sub-fractions and they form a family of similar molecules. Because of this similarity it has proved difficult to separate the sub-fractions and so far only one sequence, for rabbit thymus fraction 3 (RTL3), is available. This is given in Fig. 1.8. The N-terminal regions of some of the other subfractions have been sequenced and a comparison of 1–73 of RTL3 with RTL4 shows that 10% of the residues are different. Also included in Fig. 1.8 is the sequence of trout H1. A comparison of these sequences shows many differences in the N-terminal quarter of the molecule 1–40 and in the C-terminal half from residue 116 to the end. The central region 41–112 shows few changes and most of these are conserved. Similar results have been found for a series of H1 sequences and partial sequences from sea urchins and fish. Histone H1 possesses the most asymmetrical sequence of all the histones and can be divided into three well-defined domains: the N-terminal region 1–40 contains 75% of the lysines, prolines and alanines and no apolar residues; the central region 41–112 contains 82% of the apolar residues and no prolines; the C-terminal half from 113–213 which is very basic and contains 85% of the lysines, prolines and alanines.

The sequence of goose erythrocyte histone H5 has a similar organisation to H1 except that the basic N-terminal region is much smaller and probably does not extend beyond residue 23. H5 has a central apolar region and a very basic C-terminal half of the molecule. H5 contains about 11% arginines compared to 2% for H1 with about 25% of lysine in both molecules. The higher proportion of arginine in H5 may be required for stronger binding to DNA in erythrocyte chromatin.

1.5.6 Histone Variability

As we have just discussed, sequence studies of histones H3 and H4 show that they are among the most highly conserved of all proteins so far studied. Except for post synthetic chemical modifications which will be discussed later there is little variability of histones H3 and H4 either within one cell type or between cell types and species of higher eukaryotes. The accepted point mutation rates for H3 and H4 of about 0.1–0.2 point mutations per 100 residues (PAM units) have to be compared with PAM values of 4.0 and 14.0 for cytochrome and haemoglobin respectively. From these figures it can be deduced that although globular proteins have very specific functions and a precise three-dimensional structure, many amino acid replacements can be tolerated in the sequence. In an enzyme the active site involves only a small number of residues compared to the total number of residues in the molecule, and for the enzyme to be viable it is essential that the precise geometry of the active site be maintained. Providing that this condition is met, mutations can be tolerated in the other residues which

Fig. 1.8 — Sequences of calf thymus H1 sub-fraction RTL3 and trout testis H1.

Positions	RTL3	Trout
1–20	Ac–Ser–Glu–Ala–Pro–Ala–Thr–Ala–Ala–Pro(10)–Ala–Pro–Ser–Lys–Ala–Pro–Ala–Lys–Ala–Lys(20)	Ac–Ala–Glu–Ala–Pro–Ala–Ala–Val–0–Pro(10)–Ala–Pro–Gly–Ala–Ala–Ala–Ala–Lys–Ala–Lys(20)
21–40	0–Lys–Lys–Ala–Pro–Lys–Lys–Ala–Gly–Gly(30)–Lys–Ala–Lys–Gly–Ala–Lys–Lys–Ala–Ala–Ala(40)	Pro–Lys–Lys–Ala–Ala–Lys–Pro–Ala–Gly–Pro(30)–Gly–Ala–0–0–0–Lys–0–Lys–Gly–0(40)
41–60	Gly–Pro–Pro–Val–Ser–Glu–Lys–Val–Ala–Glu(50)–Val–Ala–Lys–Lys–Ala–Lys–Lys–Asp–Arg–Gly(60)	Gly–Pro–Ala–Val–Gly–Glu–Lys–Val–Ala–Glu(50)–Val–Ala–Lys–Lys–Ala–Ser–Lys–Asp–Arg–Gly(60)
61–80	Leu–Ser–Leu–Ala–Ala–Leu–Ala–Ala–Gly–Ser(70)–Lys–Lys–Leu–Lys–Lys–Ala–Leu–Lys–Glu–Lys(80)	Val–Ser–Leu–Ala–Ala–Leu–Ala–Ala–Gly–Ser(70)–Lys–Lys–Leu–Lys–Lys–Ala–Leu–Lys–Glu–Lys(80)
81–100	Asn–Ser–Arg–Ile–Lys–Leu–Ser–Gly–Val–Ser(90)–Gly–Lys–Gly–Lys–Leu–Ser–Thr–Tyr–Lys–Thr(100)	Asn–Ser–Arg–Val–Lys–Val–Thr–Gly–Thr–Ser(90)–Gly–Thr–Gly–Lys–Gly–Gly–Gly–Tyr–Lys–Thr(100)
101–120	Lys–Gly–Thr–Gly–Ala–Ser–Ala–Ala–Lys–Ala(110)–Lys–Lys–Lys–Ala–Ala–Lys–Glu–Gly–Glu–Ala(120)	Lys–Gly–Thr–Gly–Ala–Thr–Ala–Ala–Lys–Ala(110)–Lys–Lys–Lys–Ala–Ala–Lys–Glu–Gly–Glu–Ala(120)
121–140	Lys–Pro–Lys–Pro–0–Lys–Ala–Gly–Ala–Ala(130)–Ala–Lys–Pro–Ala–Gly–Lys–Pro–Val–Glu–0(140)	Lys–0–Lys–Ala–Ala–Lys–Ala–Ala–Pro–Ala(130)–Ala–Lys–Val–Ala–Ala–Lys–Val–Val–Ala–Lys(140)
141–160	0–Ala–Thr–Pro–Lys–Lys–Ala–Ala–Gly–0(150)–Ala–Pro–Lys–Ala–Gly–Lys–Ala–Ala–Ala–Lys(160)	Pro–Ala–Ala–Ala–Lys–Lys–Ala–Ala–0–0(150)–Ala–Pro–Ala–Ala–Ala–Lys–Ala–Ala–Ala–Lys(160)
161–180	Lys–Thr–Pro–Lys–Lys–Ala–Thr–Val–Ala–Lys(170)–Pro–Pro–Lys–Ala–Ala–Pro–Lys–Pro–Ala–Pro(180)	Lys–Ser–Ala–Lys–Lys–Ala–Ala–Val–Ala–Lys(170)–Pro–Pro–0–Ala–Ala–Pro–0–Pro–Ala–0(180)
181–200	Lys–Ser–Pro–Ala–Lys–Val–Lys–Ala–Ala–Pro(190)–Ser–Pro–Lys–Ala–0–Lys–Ala–Lys–Pro–Lys(200)	0–Thr–Pro–Lys–Lys–Ala–0–Ala–Ala–Pro(190)–Ser–Pro–Lys–Lys–Thr–Lys–Ala–0–Pro–Lys(200)
201–220	Ala–Ala–Lys–Ala–Pro–Lys–Ala–Ala–Lys–Lys(210)–Pro–0–Pro–Lys–Ala–Lys–Thr–Thr–Ala–Ala(220)	Ala–Ala–Lys–Ala–Ala–Lys–Ala–Lys–Lys–Lys(210)–Pro–Lys–Pro–Lys–Val–Lys–Ala–Ala–Pro–Ala(220)
C-term	Ala–Lys–Lys–Lys–OH	Ala–Lys–Lys–Lys–H

Fig. 1.8 Sequences of calf thymus H1 sub-fraction RTL3 (Cole R. D. and co-workers) and trout testis H1. (From Macleod A. R., Wong N. C. & Dixon G. H. (1977) *Eur. J. Biochem.* **78**, 281.)

Positions where polymorphism has been detected are underlined. Deletions are indicated by 0.

determine the overall structure of the enzyme. It follows from this that the rigid sequence conservations of histones H3 and H4 in higher eukaryotes imply that each and every residue in these proteins is involved in interactions essential to their function.

Histones H2A and H2B are more variable than H3 and H4 although with PAM values of about 1.0 they are still highly conserved compared to many globular proteins and enzymes. However the sequence variations observed for histones H2A and H2B from different sources are not distributed uniformly along the polypetide chain but are confined to the well-defined basic N- and C-terminal regions leaving the central apolar regions of these molecules rigidly conserved. It must be concluded that these conserved apolar regions, similar to histones H3 and H4, are involved in interactions essential to chromatin structure and function. It remains to be seen whether the variabilities of the basic N- and C-terminal regions are functional in that they may modulate certain aspects of chromatin structure for recognition purposes. Such considerations may be relevant to the recent findings that subsets of H2A and H2B molecules are synthesized at different stages of development of sea urchins.

H1 is the most variable of the histones with a PAM value of 4.0. As found for H2A and H2B, variations in the H1 sequence are located almost entirely in the basic N-terminal and very basic C-terminal half of the molecule leaving a conserved apolar central region. It has also been shown that there is a pattern of synthesis of H1 sub-fractions during sea urchin development. Again whether these sequence variations are functional is not known.

The properties of histones described above are summarized in Fig. 1.9. This shows the sequence asymmetries and the variable and conserved regions of histone sequences.

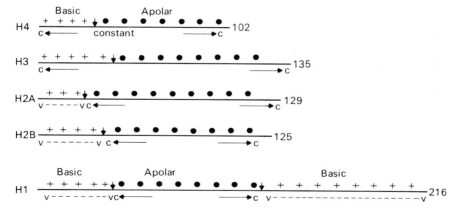

Fig. 1.9 Schematic diagram showing the properties of histones. c = constant regions and v = variable regions.

1.6 CHEMICAL MODIFICATIONS OF HISTONES

Although histone sequences, particularly of H3 and H4, are highly conserved, these same sequences are subjected to a wide variety of enzyme-controlled, post-synthetic, chemical modifications of the side chains of certain residues: acetylation and methylation of lysine residues; acetylation of N-terminal serines; phosphorylation of serines and threonines; adenosine diphosphoribosylation of glutamic acid residues. These modifications effect considerable changes in the properties of the residues. Thus the constraints which evolution has placed on histone sequences to preserve interactions essentials to certain aspects of chromatin structure and function have to be relaxed in response to other structural and functional demands of chromosomes through the cell cycle.

1.6.1 Methylation of Histones
H3 and H4 are the only histones subjected to methylation *in vivo*; in H4 the major methylation site is lysine 20 while in H3 lysines 9 and 27 are major sites and lysine 4 is a minor site. No turnover of labelled methyl groups has been detected and this modification is thought to be irreversible. It is to be noted that the sites of methylation are in the basic N-terminal regions of H3 and H4. The effect of methylation is to change the basicity and hydrophobic nature of the lysine side chain but not the net charge (Fig. 1.10(a)).

1.6.2 Acetylation of Histones
There are two types of histone acetylation; firstly the N-terminal serines of histones H1, H2A and H4 are acetylated. These acetylations are irreversible and have been associated with histone synthesis. Secondly, major reversible acetylations of lysines have been observed for histones H2A, H2B, H3 and H4 in a wide range of cells and tissues. Sites of acetylation found for trout testis histones are H4 (lysines 5, 8, 12 and 16), H3 (lysines 9, 14, 18 and 23), H2A (lysine 5) and H2B (lysines 5, 10, 13 and 19). All of the sites of acetylation are located in the basic N-terminal regions. The effect of acetylation is to convert the basic lysine side chain to a neutral acetyl lysine (Fig. 1.10(b)) and thus reduces the net basic charge of the N-terminal regions of the histones. It has been reported that multiple acetylation of histones occurs mainly in S-phase and has a rapid turnover and this has led to suggestions that acetylation of histones is involved in DNA synthesis. There are strong correlations between histone acetylation and the expression of active genes. These correlations will be discussed in more detail later in Chapter 4.

1.6.3 Histone Phosphorylation
Histone phosphorylation is a major reversible chemical modification which converts serines and threonines from neutral to negatively charged residues

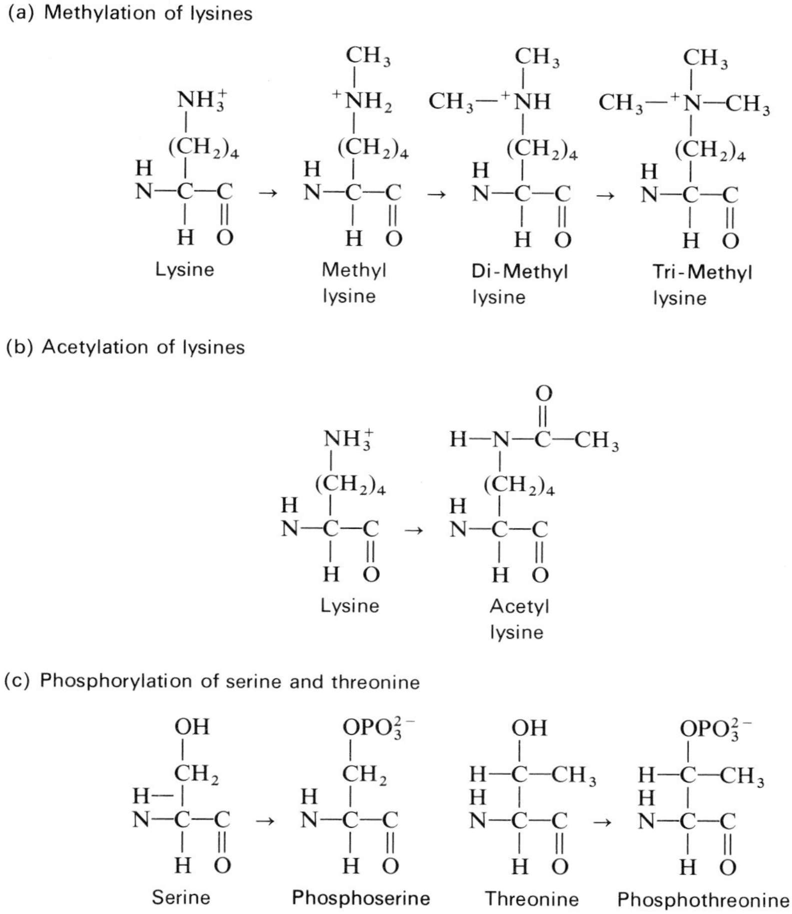

(a) Methylation of lysines

(b) Acetylation of lysines

(c) Phosphorylation of serine and threonine

Fig. 1.10 Chemical modifications of histones.

(Fig. 1.10(c). H1 is the most extensively phosphorylated of the histones and three types of phosphorylation have been identified; (1) cyclic AMP dependent phosphorylation of serine 37 which is found in the sequences of some of the H1 sub-fractions, e.g. in the sequence of sub-fraction RTL3 given in Fig. 1.8 serine 37 has been replaced by an alanine. This phosphorylation would therefore affect only those regions of chromatin containing H1 sub-fractions with a serine at position 37; (2) an *in vitro* phosphorylation of serine 105 by histone kinase HK2. This phosphorylation has not been observed *in vivo*; and (3) extensive phosphorylation of serines and threonines through the cell cycle which has been called growth associated phosphorylation. Six or seven serines and threonines are phosphorylated and so far four sites have been identified; these are threonine 16, threonine 136, threonine 153, and serine 180 which are located entirely in the basic N- and C-terminal regions of H1. Sites of phosphorylation so far identified on the other histones are; H2A (serines 1 and 19) H2B (serines 6, 14 and 36), H3 (serines 10 and 28) and on H4 (serine 1).

It can be seen from the above discussions that all of the sites of reversible modifications, acetylation and phosphorylation are located in the basic N-terminal regions of histones H2A, H2B, H3 and H4 and in the basic N- and C-terminal regions of H1. Through reducing the net positive charge of these regions both types of modification are thought to modulate histone–DNA interactions in chromatin and be involved in the conformational transitions required for DNA processing and chromosome condensation.

1.6.4 Protein A24

A24 is a very unusual hybrid protein and consists of histone H2A, (Fig. 1.7) to which a non-histone protein ubiquitin is covalently bound through the side chain amino group of lysine 119 as shown in Fig. 1.11. Ubiquitin will be discussed later. About 1 in 10 of the H2A molecules is in the form of A24 and there have been suggestions that it is involved in higher order chromatin structures, in inactive chromatin or is part of a specialized region of chromatin.

Ubiquitin — X — Gly — Gly —

Histone H2A — Lys — Lys — Thr — Glu — Ser — His — His — Lys — Ala — Lys — Gly — Lys — COOH
                       119   129

Fig. 1.11 Attachment of ubiquitin to histone H2A in protein A24. The carboxy terminus of ubiquitin is attached through Gly–Gly and an isopeptide linkage to the ε-NH$_2$ of lysine.

1.7 HISTONE COMPLEXES

A major advance in our understanding of histones came with the discovery of specific histone complexes. The first complex to be observed was an (H2A, H2B) dimer from an equimolar mixture of the two individual histones. Total histone, dissociated from DNA by exposing chromatin to high ionic strengths, can be gently separated into the components H1 (H2A + H2B)(H3 + H4). Cross-linking histones (H3 + H4) with dimethylsuberimidate (DMS) led to the identification of a tetramer (H3, H4)$_2$. DMS has the chemical structure

$$\underset{\text{MeO}-\overset{\overset{\displaystyle NH}{|}}{C}-(CH_2)_6-\overset{\overset{\displaystyle NH}{|}}{C}-OMe}{}$$

and cross-links amino groups in histones by the functional groups at the ends of the DMS molecule. From systematic studies of the cross-interactions of pairs of histones it has been shown that the strongest complexes were firstly the (H3, H4)$_2$ tetramer and then the (H2A, H2B) dimer, together with other weaker complexes. A scheme for the histone complexes is given in Fig. 1.12. By using DMS to cross-link the histone complex isolated from chromatin at 2M sodium chloride an octamer has been identified of histones consistent with the form (H2A, H2B, H3, H4)$_2$. As we shall see in the next chapter this octamer is an important component of chromatin structure.

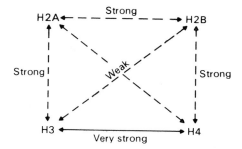

Fig. 1.12 Scheme for cross-interaction of histones as proposed by D'Anna and Isenberg.

1.8 INTERACTIONS AND CONFORMATIONS OF HISTONES

1.8.1 H2A, H2B, H3 and H4

The asymmetries in histone sequences described earlier led to the proposal that the basic N-terminal regions of the histones were required for DNA binding while the apolar regions are required for other functions. Detailed studies of the conformations and interactions of histones have given some support to this general proposal. Individual histones in aqueous solution are largely in the random coil conformation. This contrasts with globular proteins which are stabilized as globular structures in aqueous solution. With the individual histones H2A, H2B, H3 and H4 however, if the ionic strength of the aqueous solution is increased there is a conformational transition of the central and C-terminal apolar regions of the histones from the random coil to a structured state. There is evidence to suggest that these structured apolar regions then self-interact to form very large complexes, leaving the basic N- and C-terminal regions in the random coil conformation.

With the discovery of the specific histone complexes (H3, H4)$_2$ and (H2A, H2B) the question arose as to the nature of the interactions holding these complexes together. Detailed spectroscopic studies of these complexes and of complexes formed between large peptides of histones have shown that they contain two types of structure, structured central apolar regions and non-structured or random coil, basic N- and C-terminal regions. From such data, models were proposed (Fig. 1.13) in which histone complexes were held together by interactions between structured apolar regions. In the absence of DNA it appears that these basic regions are not structured. Similar behaviours have

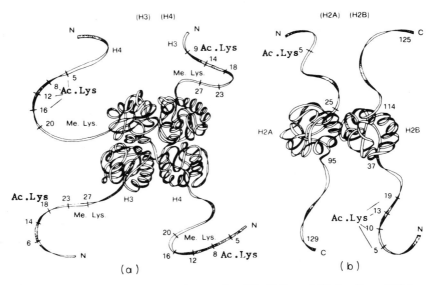

Fig. 1.13   Models for (a) the tetrameric complex (H3, H4)$_2$ and (b) the dimer (H2A, H2B). The sites of methylation and acetylation in all four histones are indicated.

Fig. 1.14   Model for the structure of free H1 relating the different structural domains with sequence variability. Sites of phosphorylation are located in the variable regions.

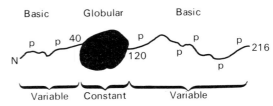

Structure of isolated histone H1 in solution.

been found for other nucleic acid binding proteins. The conserved and variable regions of histone H2A and H2B correlate closely with the structured and non-structured regions of H2A and H2B. The sites of chemical modification which are located entirely in the basic N-terminal regions of all four histones are indicated in Fig. 1.13.

1.8.2   Histone H1

Histone H1 was found to behave differently to the other histones in that it did not form large complexes on increase of ionic strength. As with the other histones, H1 is in the random coil conformation in aqueous solution. Increase in ionic strength led to a conformational transition of the central apolar region from residue 41 to 123 to a globular structure leaving the basic N- and C-terminal regions mobile and in the random coil conformation, Fig. 1.14. This structure has been confirmed by detailed studies of large peptides of H1. Also included in Fig. 1.14 are the conserved and variable regions of the H1 molecule; it is the latter regions which are phosphorylated. It can be seen that there is a clear correlation between the conserved region of H1 and the structured globular region of the molecule.

1.9 PROTAMINES

In the sperm of certain animals histones are replaced by a group of very basic proteins called protamines. Protamines have been found in many but not all fish species. They are present in herring (clupeine), salmon (salmine), trout (iridine) and tune (thynine) but not in goldfish or carp. In the last two fish species, sperm histones are present. In mammalian species protamines have been found in the sperm of mouse, bull, boar, ram, stallion and human. In the avian species protamines have been found in the rooster. The sequences of several protamines are given in Fig. 1.15. It can be seen that the fish protamines exhibit considerable homology with groupings of 4–6 arginines spaced along the molecule. The regions between the stretches of arginines contain many residues with short sidechains. These non-basic regions are thought to provide bends to allow the arginine stretches to interact favorably with DNA. In contrast, however, the sequence of the bull protamine is quite different to the fish sequences and contains one major arginine-rich region in the centre of the molecule with less basic regions at each end. Clearly a mode of interaction of this protamine with DNA would be expected to be different to that of the fish protamines.

The process when histones are replaced by protamines is called spermiogenesis and it has been studied in trout and in mice. In mice in the initial stage the primary spermatocytes synthesize variants of the histones H1, H2B and H3 which then replace the homologous somatic histones. The histones are then acetylated and replaced by a sperm-specific lysine-rich histone. Finally this lysine-rich histone is replaced by the arginine-rich protamine in the native sperm. The nucleoprotamine complex is more compact than the chromatin in somatic cells and this form is thought to provide added protection to the DNA. The structure of nucleoprotamine is quite different to chromatin and spermiogenesis provides an interesting process whereby, through histone modification, sperm specific histones and the interaction of protamines, the structural changes are accomplished.

1.10 NON-HISTONE PROTEINS

Excluding histones, all other proteins associated with eukaryotic chromosomes are grouped together under the general name of non-histone proteins (NHP).

Fish protamines

CLUPEINE Y1	Ala	Arg Arg Arg Arg	Ser Ser Ser Arg Pro Ile	Arg Arg Arg Arg	Pro Arg Arg Arg	Thr Thr	Arg Arg Arg Arg	Ala Gly	Arg Arg Arg Arg
SALMINE A1	Pro	Arg Arg Arg Arg	Ser Ser Arg Pro Val	Arg Arg Arg Arg Arg	Pro Arg	Val Ser	Arg Arg Arg Arg Arg Arg	Gly Gly	Arg Arg Arg Arg
IRIDINE I1	Pro	Arg Arg Arg Arg	Ser Ser er Arg Pro Val	Arg Arg Arg Arg	Ala Arg Arg	Val Ser	Arg Arg Arg Arg Arg Arg	Gly Gly	Arg Arg Arg Arg
THYNNINE Y1	Pro	Arg Arg Arg Arg	Glu Ala Ser Arg Pro Val	Arg Arg Arg Arg Arg Arg	Tyr Arg Arg Ser Thr Ala Ala	Val Val	Arg Arg Arg Arg Arg Arg	Val Val	Arg Arg Arg Arg

Bull protamine

Ala Arg Tyr Arg Cys Cys Leu Thr His Ser Gly Ser Arg Cys Arg Arg Arg Arg Arg Cys Arg Arg Arg Arg Arg Arg Pro Gly Arg Arg Arg Arg Arg Arg Val Cys Tyr Thr Val Leu Arg Cys Thr Arg

Fig. 1.15 Protamine sequences.

Figure 1.5 shows some NHP from chicken erythrocyte nuclei. NHP have also been called 'acidic proteins' to distinguish them from the basic histones. The NHP include the enzymes required for DNA replication, enzymes required for the histone modifications described above and a whole multitude of enzymes required for functions of the cell-nucleus, i.e. the 'housekeeping' enzymes. Until recently the analysis of the NHP of the eukaryotic cell nucleus was an intractable problem. SDS-polyacrylamide gel electrophoresis separated the components of the NHP into about 100 bands and similar patterns were observed for the NHP of the different tissues from a single organism. This showed that the major components of the NHP were the same in the different tissues and were thus mainly the proteins involved in metabolic processes common to all cells. To observe differences in the NHP components of different cells clearly required much higher resolution in protein separation and identification than was available at that time. Why go to the considerable lengths required to undertake a detailed analysis of the NHP? The reason is that some understanding of the control of gene expression must come from the identification and separation of the proteins involved in these processes. A major difficulty with their identification results from the small numbers of copies of regulatory proteins (of the order of hundreds) expected to be present in the different cells.

The technical difficulties in studying this problem have been greatly reduced by the introduction of two-dimensional polyacrylamide gel electrophoresis which has the resolution to separate thousands of proteins. This technique has been applied to the differentiation of a single cell type—the Friend leukemic cell which can be transformed to an erythropoietic cell by exposure to dimethyl-sulphoxide (DMSO). It is not known how DMSO causes this effect. However, addition of DMSO to the culture medium of Friend cells induces the cells to produce haemoglobin and after 4–5 days this protein constitutes 20–25% of the cell protein. The total cell content of proteins, cytoplasmic and nuclear, of Friend cells before and after transformation are shown in Fig. 1.16. As can be seen, except for about five differences, the total protein patterns are surprisingly similar. Although not proven, it is thought that the globin gene regulatory proteins are possibly included in the small number of observed differences. The possibility, however, cannot be excluded that some of these important proteins are present in amounts too small to be observed even on the 2-D (two-dimensional) gels.

It should not be thought that histones are the only structural proteins in chromosomes. Other structural proteins are included in the NHP. It has recently been shown that a group of NHP, called 'scaffold' proteins, are involved in determining the shape of the metaphase chromosome and are also thought to be involved in organizing metaphase chromosomes. This will be discussed in the next chapter. Chromatin can take up different conformational states. This is clearly demonstrated by the series of structural changes which are observed

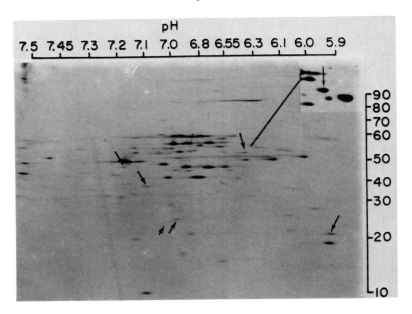

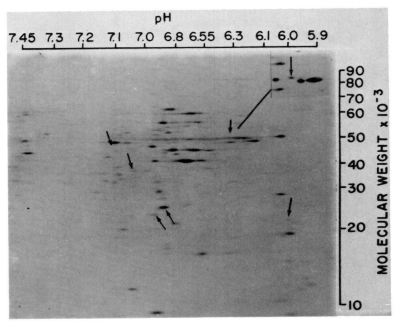

(a)

Fig. 1.16 Two-dimensional electrophoresis gels of total protein from Friend cells before and after transformation with DMSO. The protein had been previously labelled with radioactive methionine and the smears on the gel are the results of autoradiographic exposure. The upper gels are of cells untreated with DMSO, the lower are from

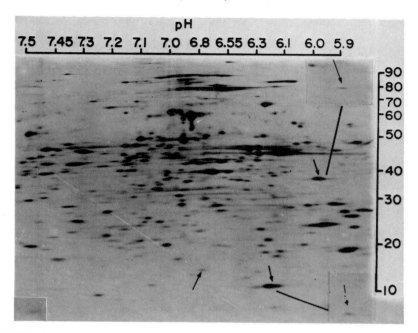

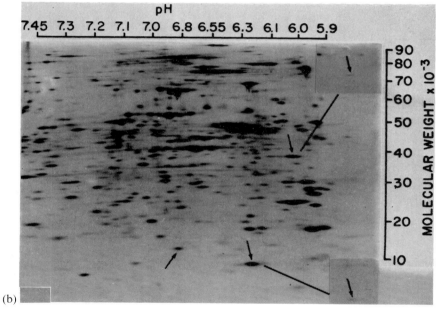

(b)

treated cells. The (a) gels have been exposed photographically for a shorter time than (b). Arrows indicate proteins which are present in very different amounts in (a) & (b). (From Peterson J. L. & McConkey E. H. (1976) *J. Biol. Chem.* **251**. 555–8.)

during chromosome condensation to the metaphase chromosome. There is evidence also to show that 'active' chromatin, i.e. regions which are active in transcription, is structurally different to 'inactive' chromatin. Although chemical modifications of histones have been implicated in these processes there is now evidence which suggests the involvement of another group of NHP, called, because of their mobilities on acrylamide gels, high mobility group (HMG) proteins.

1.11 HMG PROTEINS

HMG proteins were first identified by Johns and co-workers in calf thymus chromatin. So far four of these proteins have been characterized; HMG 1, 2, 14 and 17. Two HMG proteins have also been characterized from trout testis; HMGT, similar to HMG 1 and 2, and H6, similar to HMG 14 and 17. It was initially thought that there were a large number of HMG proteins but it has now been shown that many of these were proteolytic degradation products of other chromosomal proteins. There are approximately 10^6 molecules of each HMG protein in the cell nucleus, which is too large a number for them to be considered as candidates for gene regulatory proteins. Their unusual properties suggest a structural role in chromatin.

1.11.1 Properties of HMG Proteins

The four calf thymus HMG proteins comprise two pairs of very similar proteins, HMG 1 and 2, and HMG 14 and 17. Both HMG 1 and 2 proteins have about 250 residues, similar amino acid sequences and structural properties. HMG 14 and 17 have 100 and 89 amino acids respectively and they also have similar sequences and conformations. Unusual features of the amino acid compositions of these proteins (Table 1.2) are the high contents of basic residues, 25–28 % and also of acidic residues 22–29 %, thus giving them overall between 51 and 54 % of charged residues. HMG 14 and 17 also contain high proportions of helix destabilizing residues, 22.5 and 26.4 % respectively of proline, glycine and serine taken together and very few apolar residues, leucine, isoleucine and valine. These compositions can be related to their conformation behaviour.

HMG proteins have been found in all tissues and higher organisms so far examined. A range of tissues from calf thymus, kidney, liver and spleen contain identical complements of the four HMG proteins. Similar HMG proteins have been found in avian tissues though their quantities are thought to be less than for other tissues. In trout testis two HMG proteins have been found, HMGT and H6. H6 was originally thought to be a histone because of its high lysine content and this overlap of nomenclature draws attention to some of the similarities between HMG proteins and histones. Trout liver and spleen contain, in addition to HMGT and H6, two proteins similar to HMG 14 and 17. As for

Table 1.2 Amino acid analyses (moles %) of calf thymus HMG proteins. (Table provided by Dr E. W. Johns.)

Amino acid	HMG 1	HMG 2	HMG 14	HMG 17
Asp	10.7	9.3	8.3	12.0
Thr	2.5	2.7	4.1	1.2
Ser	5.0	7.4	8.0	2.3
Glu	18.1	17.5	17.5	10.5
Pro	7.0	8.9	8.1	12.9
Gly	5.3	6.5	6.4	11.2
Ala	9.0	8.1	14.8	18.4
Val	1.9	2.3	4.0	2.0
$\frac{1}{2}$-Cys	Trace	Trace	—	—
Met	1.5	0.4	—	—
Ile	1.8	1.3	—	—
Leu	2.2	2.0	2.0	1.0
Tyr	2.9	2.0	—	—
Phe	3.6	3.0	—	—
Lys	21.3	19.4	21.1	24.3
His	1.7	2.0	—	—
Arg	3.9	4.7	5.4	4.1

lysine-rich histones, microheterogeneity has been observed for HMG proteins; in particular for HMG 2 which has four sub-fractions. Chemical modifications (phosphorylation and acetylations) of the HMG proteins have been reported and this may account for some of their microheterogeneities.

1.11.2 Sequences of HMG Proteins

HMG 14

The sequence of calf thymus HMG 14 is given in Fig. 1.17. It can be seen that there is an asymmetry in the distribution of basic and acidic residues; the N-terminal 61 residues contain 22 basic residues (i.e. 36%) and 6 acidic residues (10%) while the C-terminal 40 residues contain 4 basic residues (10%) and 14 acidic residues (35%). The first 33 residues of HMG 14 from chicken erythrocytes have been sequenced and show 6 differences to the calf thymus sequence; from the amino acid composition of chicken erythrocyte HMG 14 there is expected to be about 11% variation compared to the calf protein. This is a much larger variation than found for the homologous histones from the two tissues.

HMG 17

The sequence of calf thymus HMG 17 is given in Fig. 1.18. The N-terminal region is basic; in the first 60 residues there are 22 lysines and arginines (i.e. 37%) while in the C-terminal 30 residues there are only 3 (10%). A feature of this sequence, not found in other HMG proteins and histones, is the presence of 6 prolines in the centre of the molecule from residue 31 to 40.

Pro — Lys — Arg — Lys — Val — Ser — Ser — Ala — Glu — Gly[10] —

Ala — Ala — Lys — Glu — Glu — Pro — Lys — Arg — Arg — Ser[20] —

Ala — Arg — Leu — Ser — Ala — Lys — Pro — Ala — Pro — Ala[30] —

Lys — Val — Glu — Thr — Lys — Pro — Lys — Lys — Ala — Ala[40] —

Gly — Lys — Asp — Lys — Ser — Ser — Asp — Lys — Lys — Val[50] —

Gln — Thr — Lys — Gly — Lys — Arg — Gly — Ala — Lys — Gly[60] —

Lys — Gln — Ala — Glu — Val — Ala — Asn — Gln — Glu — Thr[70] —

Lys — Glu — Asp — Leu — Pro — Ala — Glu — Asn — Gly — Glu[80] —

Thr — Lys — Asn — Glu — Glu — Ser — Pro — Ala — Ser — Asp[90] —

Glu — Ala — Glu — Glu — Lys — Glu — Ala — Lys — Ser — Asp[100] —

Fig. 1.17   Sequence of cal thymus HMG 14. (From Walker J. M. *et al.* (1979) *F.E.B.S. Lett.* **100**, 394.)

Pro — Lys — Arg — Lys — Ala — Glu — Gly — Asp — Ala — Lys[10] —

Gly — Asp — Lys — Ala — Lys — Val — Lys — Asp — Glu — Pro[20] —

Gln — Arg — Arg — Ser — Ala — Arg — Leu — Ser — Ala — Lys[30] —

Pro — Ala — Pro — Pro — Lys — Pro — Glu — Pro — Lys — Pro[40] —

Lys — Lys — Ala — Ala — Pro — Lys — Lys — Set — Glu — Lys[50] —

Val — Pro — Lys — Gly — Lys — Lys — Gly — Lys — Ala — Asp[60] —

Ala — Gly — Lys — Glu — Gly — Ash — Asn — Pro — Ala — Glu[70] —

Asn — Gly — Asp — Ala — Lys — Thr — Asn — Gln — Ala — Glu[80] —

Lys — Ala — Glu — Gly — Ala — Gly — Asp — Ala — Lys —

Fig. 1.18   Sequence of calf thymus HMG 17. (From Walker J. M. *et al.* (1980) *F.E.B.S. Lett.* **112**, 207.)

HMG 1 and 2

Although HMG 1 and 2 have not been fully sequenced there is sufficient data from the partical sequences and amino acid compositions to discuss the properties of the primary sequences. From this data it can be seen that HMG 1 and 2 have similar sequences. A very unusual feature of both proteins is the presence of an unbroken run of 40 aspartic and glutamic acid residues in the C-terminal regions of the molecules. Its function is not understood. The N-terminal regions of these proteins contain a high proportion of basic and apolar residues. Both HMG 1 and 2 are therefore asymmetric with basic N-terminal regions and acidic C-terminal regions. It has also been shown that 2–3% of the lysine residues in both proteins are acetylated.

1.11.3 Sequence Homologies of HMG Proteins

A comparison of the sequences of HMG 14, 17 and trout H6 show that there is extensive homology between the basic N-terminal halves of these proteins. This homology breaks down in the centres of the molecules and very limited homology is found from the C-terminal regions. Some homologies have also been found between the N-terminal regions of HMG 14 and 17 and the basic N-terminal regions of histones particularly with the very lysine-rich histone H1.

1.12 CONFORMATIONAL STUDIES OF HMG PROTEINS

Spectroscopic and other physical studies show that HMG 1 and 2 contain 40–50% α-helix and exhibit many of the properties of globular structured proteins. In contrast similar studies of HMG 14 and 17 show that they adopt the random coil conformation in aqueous solution over a wide range of ionic strengths. These conformational behaviours are consistent with the properties and partial sequence data of HMG 1 and 2 and with the sequences of HMG 14 and 17. The latter proteins contain very few apolar residues but high proportions of charged residues and helix destabilizing residues such as proline and glycine.

1.13 INTERACTIONS OF HMG PROTEINS WITH DNA

All four HMG proteins bind to DNA. The nature of the interaction is primarily ionic, involving basic residues and the DNA phosphate groups. Physical studies of the interaction of HMG 17 with DNA show that the basic and proline rich region between residues 15 and 40 is a major binding site. HMG 1 and 2 appear also to bind to double stranded DNA through the basic N-terminal region leaving the run of 40 acidic residues in the C-terminal region uncomplexed. There is some evidence to suggest that HMG 1 and 2 may be DNA-unwinding proteins. Complexing these proteins to closed circular DNA results in a change in the linkage number by unwinding the DNA double helix. HMG 14 and 17 do not have this effect.

1.14   UBIQUITIN

The name ubiquitin comes from the widespread occurrence of this protein in all cells, from both prokaryotes and eukaryotes so far examined. From this it appears that ubiquitin has essential but as yet unknown functions in all living cells. In mammalian tissues ubiquitin is present to the amount of 2–3 % of one of the conserved histones. The discovery of the bifurcated A24 protein (Fig. 1.11) consisting of ubiquitin covalently bound to histone H2A was the first direct evidence of a role for ubiquitin in chromatin structure and function. Since then, free ubiquitin has been found in calf thymus and trout testis chromatin and an inverse correlation has been reported of the amounts of ubiquitin and A24 in active and inactive chromatin; free ubiquitin being found largely in active chromatin and A24 largely in inactive chromatin.

The sequence of calf thymus ubiquitin (Fig. 1.19), is typical of a globular protein with a high proportion of apolar residues distributed uniformly throughout the molecule and physical studies have demonstrated that ubiquitin is a very stable globular protein. The sequence of ubiquitin is as highly conserved as histones H3 and H4; trout testis ubiquitin is identical to that from calf thymus for at least 71 of the 74 residues.

$$
\begin{array}{llllllllll}
\text{Met} - \text{Gln} - \text{Ile} - \text{Phe} - \text{Val} - \text{Lys} - \text{Thr} - \text{Leu} - \text{Thr} - \overset{10}{\text{Gly}} - \\
\text{Lys} - \text{Thr} - \text{Ile} - \text{Thr} - \text{Leu} - \text{Glu} - \text{Val} - \text{Glu} - \text{Pro} - \overset{20}{\text{Ser}} - \\
\text{Asp} - \text{Thr} - \text{Ile} - \text{Glu} - \text{Asn} - \text{Val} - \text{Lys} - \text{Ala} - \text{Lys} - \overset{30}{\text{Ile}} - \\
\text{Gln} - \text{Asp} - \text{Lys} - \text{Glu} - \text{Gly} - \text{Ile} - \text{Pro} - \text{Pro} - \text{Asp} - \overset{40}{\text{Gln}} - \\
\text{Gln} - \text{Arg} - \text{Leu} - \text{Ile} - \text{Phe} - \text{Ala} - \text{Gly} - \text{Lys} - \text{Gln} - \overset{50}{\text{Leu}} - \\
\text{Glu} - \text{Asp} - \text{Gly} - \text{Arg} - \text{Thr} - \text{Leu} - \text{Ser} - \text{Asp} - \text{Tyr} - \overset{60}{\text{Asn}} - \\
\text{Ile} - \text{Gln} - \text{Lys} - \text{Glu} - \text{Ser} - \text{Thr} - \text{Leu} - \text{His} - \text{Leu} - \overset{70}{\text{Val}} - \\
\text{Leu} - \text{Arg} - \text{Leu} - \text{Arg} \\
\end{array}
$$

Fig. 1.19   Sequence of calf thymus ubiquitin (Schlesinger D. H. *et al.* (1975) *Biochemistry* **14**, 2214).

1.15   FURTHER READING
See end of Chapter 2.

Chapter 2
Chromatin Structure

2.1 INTRODUCTION

There are major problems in understanding the structure of chromatin and chromosomes which result from the extraordinary lengths of DNA in the genome. As mentioned earlier it is generally accepted that only 5–10% of this DNA is used in coding for proteins and various RNA molecules, which raises major questions as to the functions of the very large amounts of non-coding DNA, e.g. is part of this DNA used in the structural organization of the chromosomes? Eukaryotic chromosomes are very complex and it is to be expected that rules exist for the control of chromatin structure in response to the requirements of cells and organisms. In progressing through the cell cycle chromosomes undergo a series of conformational changes which involve all of the components of the chromosome; DNA, histones and components of the NHP. It is reasonable to expect, therefore, that specific regions of the DNA may be involved in the structure and organization of the chromosome. During differentiation, mechanisms, as yet unknown, select out those genes which are to be active in a particular tissue (see discussion in Chapter 7). The active genes are only a very small proportion of the total number of genes in the organism and probably have a different conformation to the large number of inactive genes (see pages 200 and 251). In addition to structural differences mechanisms must also exist for controlling their genetic activity in response to the demands of the cell. We are therefore concerned with: (1) the structure of inactive chromatin; (2) the structure of active genes; (3) the different conformational states of chromosomes as they go through the cell cycle and the control of these events.

2.2 SUB-UNIT STRUCTURE OF CHROMATIN

Until recently the basic molecular structure of chromatin was thought to be a regular 'supercoil' in which the DNA was coiled by interactions with histones into a uniform hollow coil with a diameter of 11 nm and a pitch of 12 nm. This view was changed by the seminal observations of Hewish and Burgoyne on the effect of an endonuclease on chromatin in rat liver nuclei. The endonuclease which cleaves double stranded DNA was activated by the addition of the divalent cation Ca^{2+} and allowed to digest rat liver chromatin. Gel analysis of the DNA digestion products gave results similar to those shown in Fig. 2.1(b) for

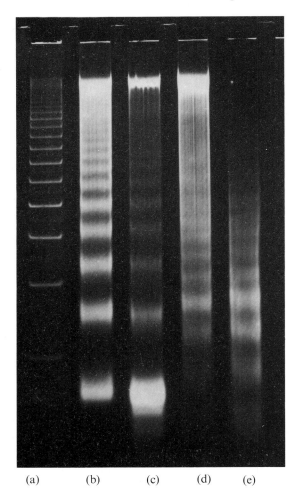

(a)   (b)   (c)   (d)   (e)

Fig. 2.1 Digestion of mouse liver nuclei with micrococcal nuclease (slots b and c) and DNase II (slots d and e). Incubations were for 1h at 37 °C. In (b) and (c) micrococcal nuclease was added at concentrations of 7 and 25 U ml^{-1} respectively. In (d) and (e) DNase II was added at concentrations of 125 and 250 U ml^{-1} respectively. Slot (a) contains a mouse satellite DNA marker partially digested with *Eco* RII. (Altenburger F. R. M. *et al.* (1976) *Nature* **264**, 517.)

micrococcal nuclease digestion of mouse liver nuclei i.e. a series of bands of discrete DNA lengths in which the lengths of the DNA in the higher members of the series were integral multiples of the smallest DNA length shown later to be about 200 bp of DNA. These observations led to the proposal that chromatin consisted of a repeating sub-unit structure in which the DNA linking the sub-units was accessible to nuclease attack. In parallel with the use of an endo-nuclease as a biochemical structural probe, chromatin subunits were visualized in electron micrographs by Olins and Olins and also by Woodwock which showed that gently prepared chromatin had a 'beaded' appearance (Fig. 2.2). A bead, about 11–12 nm in diameter, was equated with the chromatin subunit deduced from the nuclease digestion results.

Our understanding of chromatin structure has been revolutionized by results obtained from the systematic applicating of the different nucleases to chromatin structure. In addition to their use as a biochemical probe of chromatin structure they have been used to prepare large quantities of chromatin sub-units for physical studies. The full chromatin sub-unit is now called the 'nucleosome'.

2.3 THE NUCLEOSOME

2.3.1 Composition of the Nucleosome

The DNA lengths in nucleosomes from many, but not all, tissues are about 195 ± 5 bp (in this chapter we will often refer to this as being 200 bp). From the observation of the specific histone complex (H3, H4)$_2$ described in the previous chapter Kornberg proposed that the nucleosome contained 200 bp DNA together with two copies of the histones H2A, H2B, H3 and H4 and one copy of the very lysine-rich histone H1. Cross-linking of histones in the nucleosome showed the presence of an octamer (H2A, H2B, H3, H4)$_2$. Although there have been some doubts concerning the exact number of H1 molecules per nucleosome because of difficulties of biochemical isolation, it is now thought that there is normally one H1 molecule associated with each nucleosome.

2.3.2 Variability in DNA Content of Nucleosomes

A survey of the DNA content of nucleosomes from different tissues and organisms is shown in Table 2.1. It can be seen that there is a wide range of values. Some general features emerge; firstly, the smallest DNA repeats are found for the lower eukaryotes, *Aspergillus*, yeasts, etc; secondly, nucleosomes from most somatic tissues have a DNA content of 195 ± 5 bp DNA. There are however notable exceptions to this, particularly for those cells which are known to contain different lysine-rich histones. An obvious example is provided by avian erythrocytes in which H1 is largely replaced by another very lysine-rich histone, H5. The DNA content of the chicken erythrocyte nucleosome is 212 bp. The longest DNA repeat so far is for the sea urchin sperm nucleosomes which contain 241 bp whereas the sea urchin gastrula nucleosome has a DNA content of 218 bp. An interesting difference has been observed for two brain cells; nucleosomes from genetically active rabbit cortical neuronal cells have a DNA content of 162 bp whereas for those from the relatively inactive rabbit cortical glial cells it is 197 bp. It has been reported that the histones from the more active neuronal cells are acetylated. In general it is thought that the differences in the DNA contents given in Table 2.1 are correlated in some way with differences in the histones, with different types of H1, H2A and H2B histones and differences in the states of chemical modifications of the histones. It is unlikely that the DNA content of nucleosomes within even one cell type is constant. Many of the DNA repeats given in Table 2.1 are for a single nuclease digestion time whereas to investigate this question fully requires a time depend-

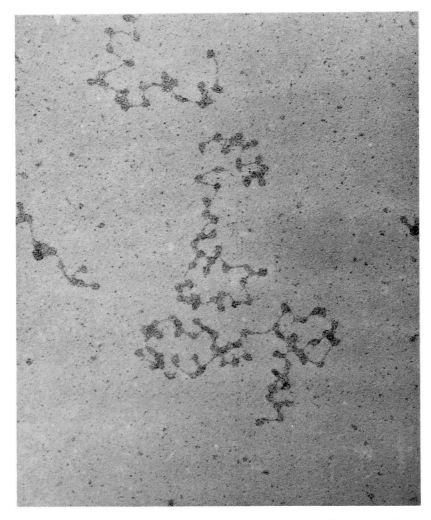

Fig. 2.2 Electron micrograph of an extended fragment of mouse liver chromatin.
(Courtesy Miller F., Igo-Kemenes T. and Zachau H. G.)

ence study. The two DNA repeats given for the true slime mould *Physarum polycephalum* are the result of such a study. For short times of nuclease digestion the DNA repeat is 191 bp DNA. With increasing times of digestion the repeat length is found to be reduced in length reaching a minimum repeat of 171 bp DNA. Similar results have been found for yeast and suggests that the more accessible regions of chromatin have larger DNA repeats while less accessible regions have shorter DNA repeats. Correlations are now being made between these more accessible regions and the genetically active regions of the genome.

Table 2.1 DNA Content of Nucleosomes (data from Compton J. L. *et al.* (1976) *Proc. Natl. Acad. Sci.* **73**, 4382)

Cell Type	DNA Content of Nucleosome (base pairs)
Aspergillus	154
Yeast	165, 163
Neurospora	170
Physarum	171, 190
Rabbit Cortical Neuron	162
Cells Grown in Culture:	
CHO	177
HeLa	183, 188
P815	188
Myoblast	189
Rat Kidney Primary Culture	191
Rabbit Cortical Glia	197
Rat Bone Marrow	192
Rat Fetal Liver	193
Rat Liver	198, 196
Rat Kidney	196
Chick Oviduct	196
Chicken Erythrocyte	207, 212
Sea Urchin Gastrula	218
Sea Urchin Sperm	241

2.3.3 Chromatin Core Particles

The ladder of DNA lengths shown in Fig. 2.1b is obtained at low concentrations of staphylococcal nuclease. At higher concentrations of nuclease the yield is largely of mononucleosome (Fig. 2.1(c)). The digestion, however, does not stop at this point and at longer times of digestion the DNA ends of the nucleosome are trimmed to give well-defined sub-nucleosomal particles. As can be seen in Fig. 2.1(c) for the mononucleosome there is a spread of DNA lengths which is centred about 190 bp DNA. With increasing time of digestion this broad band becomes sharper and moves to a 168 bp DNA length. Associated with the corresponding 168 bp chromatin particle is the histone octamer (H2A, H2B, H3, H4)$_2$ and histone H1. Further digestion reduces the DNA length to 146 bp DNA which is associated with the histone octamer but no H1. This sub-nucleosome particle is called the 'core particle'. The basic ideas concerning chromatin structure which have emerged from micrococcal nuclease digestion studies are given by the equation:

$$\text{Nucleosome} = \underbrace{146\,\text{bp DNA} + 2(\text{H2A,H2B,H3,H4})}_{\text{core particle}} + \text{Linker DNA} + \text{H1}$$

The linker DNA has a variable length depending on the origin of the nucelosome.

2.4 STRUCTURE OF THE NUCLEOSOME AND CORE PARTICLE

2.4.1 Digestion of Chromatin with DNase I and DNase II

Further information on the structure of the nucleosome has been obtained by the use of DNase I and II which are enzymes derived from eukaryotic tissues. DNase I produces single strand breaks (nicks) in the DNA, whereas DNase II, like staphylococcal nuclease, makes complete double strand breaks. The single stranded DNA products of DNase I digestion of chromatin are shown in Fig. 2.3

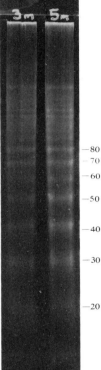

Fig. 2.3 Ladder of single stranded DNA lengths from DNase I digestion of He La cell nuclei. Although the ladder is labelled 20 bases, 40 bases etc. there is a 10.4 base interval between 20 and 145 bases.

and form a ladder of single stranded DNA lengths with a 10.4 base interval from 20 bases to greater than 145 bases. As can be seen, the intensities of the bands are not uniform, indicating variation in the frequencies of cutting at the sites of cleavage. The simple interpretation of the 10.4 base DNA ladder is that the DNA must be exposed to the action of DNase I approximately every 10 bases of DNA. If the DNA in a modified B-form were on the outside of the core particle a cleavage site would be accessible to DNase I roughly every 10 bases.

The frequency of cleavage by DNase I would also depend on the degree of protection afforded to the DNA by the interactions of the histones. We shall see later (Chapter 5) that at short times of digestion DNase I preferentially attacks DNA in actively transcribing chromatin and has been used to probe the properties of active chromatin particularly the nature of the proteins associated with active chromatin.

Digestion of chromatin by another nuclease, DNase II, was found to give a DNA digestion pattern strikingly different from that of staphylococcal nuclease (Fig. 2.1(d) & (e)). Instead of a '200 bp' DNA ladder a 100 bp DNA ladder is observed. Nevertheless after DNase II digestion full size nucleosomes are isolated but the DNA associated with the nucleosome is cut at the mid point at 100 bp. This strongly suggests that the nucleosome has two fold symmetry and and that DNase II is able to cut DNA at this axis of symmetry. The 100 bp DNA ladder is observed at higher ionic strengths when it is known that chromatin is in the second order structure. At lower ionic strengths when chromatin is in the extended form like a string of beads a 200 bp DNA ladder is obtained. DNase II therefore also probes for the higher order chromatin structure.

2.4.2 Spatial Distribution of Histones and DNA in Nucleosomes

Analysis of the scattering of radiation by biological particles in isolation gives information on the shape and size of the particle. For some years X-ray and light scattering have been used extensively to study proteins, nucleic acids, ribosomes, etc. With the advent of high flux neutron generators neutron scattering is being increasingly used and is proving to be the most powerful technique for study of the structures of two-component biological systems, such as nucleosomes, in solution. With this technique it is possible to obtain structural data on either the histone or the DNA component of the nucleosome. This ability to separate the neutron scattering effects of histone and DNA comes from the fact that the neutron scattering length of hydrogen ^{1}H *is negative* (b $= -0.378 \times 10^{-12}$ cm) while that for deuterium ^{2}H *is positive* (b $= +0.65 \times 10^{-12}$ cm). The neutron scattering lengths of the other elements found in biological macromolecules are also positive and they are of similar value; C ($+0.66 \times 10^{-12}$ cm), O ($+0.58 \times 10^{-12}$ cm), N ($+0.94 \times 10^{-12}$ cm) and P ($+0.53 \times 10^{-12}$ cm). A consequence of this difference between ^{1}H and the other elements is that the higher the proportion of hydrogen with respect to the other elements, C, O, N and P in a biological molecule, the lower is the scattering length density of the molecule. The scattering length density is the sum of the scattering lengths within a unit volume in the macromolecules. Thus the scattering length density of the histone is 2.29×10^{10} cm^{-2} while for DNA it is 3.85×10^{10} cm^{-2}. This difference can be exploited by studying the neutron scatter of nucleosomes or core particles in mixtures of light and heavy water, ^{1}H$_2$O and ^{2}H$_2$O. Because of the difference in scattering length of ^{1}H and ^{2}H the neutron scattering length density of ^{1}H$_2$O is -0.55×10^{10} cm^{-2} while that of ^2H_2O is 6.36×10^{10} cm^{-2} and by taking mixtures of ^1H_2O and ^2H_2O any value of scattering length density between these extremes can be obtained. Thus the water mixture 37% ^2H_2O:63% ^1H_2O has the same scattering length density as histone. For core particles in this water mixture the histones are matched by the solvent and the DNA component dominates the neutron scatter (Fig. 2.4(a)). At 65% ^2H_2O:35% ^1H_2O the DNA component is matched by the solvent and the histone component dominates the neutron scatter (Fig. 2.4(b)). This procedure is called contrast matching.

From the neutron scatter curves of the core particles in the above water mixtures the radius of gyration R_G of the histone and of the DNA in the particle can be obtained. The dimensions of the particle and its components can be determined from the radii of gyrations if the shape is known. However, for chromatin core particles the radius of gyration of the DNA component is 4.7–4.9 nm while that of the histone is about 3.0–3.3 nm. These values are very different and clearly demonstrate that the DNA is external to a histone core in the subunit. A full analysis of the neutron scatter curves showed that the centre of scattering mass of the histones was approximately coincident with the centre of scattering mass of the DNA and that the maximum dimension of the core particle was 11.0 nm.

2.4.3 Shape of the Nucleosome and Core Particle

It was thought initially that the shape of the nucleosome was close to spherical with a diameter about 10–11 nm. From electron microscopy studies of the isolated nucleosome, however, it was suggested that the shape of the nucleosome was more like a disc. From the neutron scatter study outlined above it is possible to obtain scatter functions which are related to the pure shape of the particle Ic and to the internal structure of the particle Is.

Such scatter functions and the scatter curves of the core particle in ^2H_2O can be used to test the various models proposed for the nucleosome and it was found that the spherical models did not fit these functions. The shape which gave the best fit to the data was an oblate spheroid of axial ratio 0.5. This ratio, taken with the maximum dimension of the particle of 11.0 nm, gives a flat disc $11.0 \times 11.0 \times 5.5$ nm. Note that for the radius of gyration of the DNA of 4.9 nm to be consistent with this shape the DNA must be located on the edge of the disc. The length of 140 bp of DNA is 47.6 nm and this corresponds to 1.7 ± 0.2 turns of DNA on the periphery of the disc. More detailed fitting of models to the shape and internal structure functions led to the proposal of a model for the core particle in solution which consisted of a histone core of shape $6.6 \times 6.6 \times 6.0$ nm with 1.7 turns of DNA coiled on the outside with a pitch or around 3.0 nm. Neutron scatter studies as described above give data on spherically averaged structures in solution. These data are extensive however and lead to the firm proposals of models for which there is little doubt about the shape or the fact that its DNA is external to a histone core. In the final analysis,

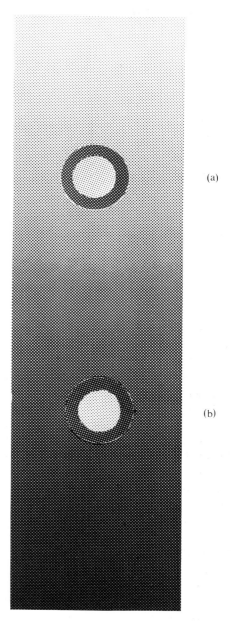

(a)

(b)

Fig. 2.4 Illustration of neutron contrast matching technique. The gradation of background represents a continuum of neutron scattering densities resulting from different mixtures of light and heavy water. Within the circles, the light centre represents histone and the dark perimeter represents DNA. The 37% ^2H_2O: 63% ^1H_2O water mixture has the same neutron scattering length diversity as histone and DNA on the outside of the particle dominates the neutron scatter. At 65% ^2H_2O:35% ^1H_2O the reverse situation holds and histone dominates the scatter.

however, solution scatter studies cannot prove that a model is correct nor give detailed models at high resolution. Such proof and atomic models can only come from X-ray or neutron diffraction studies of single crystals. Fortunately, the core particles are sufficiently regular that crystals can be grown. So far the crystals have given only limited diffraction data. These data, however, taken in conjunction with electron microscopic studies of the crystal have led to an independent proposal of a model very similar to that proposed for core particles in solution. This model from both X-ray diffraction and neutron scatter studies is given in Fig. 2.5 and has some interesting features. Firstly, it is disc shaped

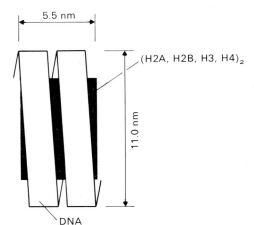

Fig. 2.5 Model for the structure of the core particle from neutron scatter solution studies and X-ray and electron microscopy studies of crystals.

and as we shall see later this leads to difficulties in understanding how the discs are arranged in higher order structures. Secondly, the pitch of the DNA coil around the histone core is very small, around 2.7 nm (X-ray) or 3.0 (neutron scatter) i.e. the DNA is tightly coiled. Since the diameter of the DNA molecule is 2.0 nm this means that the facing edges of adjacent turns of the DNA are very close together and separated by only 0.7–1.0 nm. Note that such a close approach distance would allow basic regions of histones to interact with two turns of DNA simultaneously. In relating the model to the properties of histones described in the previous chapter it is probable that the complex of apolar regions of the histones (see Fig. 1.13) comprises the apolar core of the core particles while the basic N-terminal regions of the histones are involved in other interactions. This could involve the DNA around the core particle, the linker DNA or DNA in higher order chromatin structures.

The path of the DNA around the histone core is not known precisely but this should be obtained eventually from the single crystal diffraction studies. It is however quite clear that the DNA follows a radius of about 4.5–5.0 nm and because of the constraints of the neutron data on the model, the DNA coil cannot be very different from that outlined above. In relation to the diameter of the DNA molecule of 20 nm this radius is very small and it has been suggested that a large amount of energy would be required to bend DNA to this radius. For this reason the idea of 'kinks' was introduced to alleviate the strain of bending the DNA. Thus instead of a uniform deformation of the DNA the energy of strain would be concentrated at regularly spaced points on the DNA causing kinks at those points. It was also thought that regularly spaced kinks in the DNA would give another explanation for the 10 bp DNA ladder (Fig. 2.3) obtained from DNase I digestion of chromatin, i.e. that the DNase I would cleave in the regions of the exposed kinks. There is however no evidence to support the idea of kinks in DNA coiled around the periphery of the core particle disc and recent calculations of the energies required to deform DNA have shown that such energies are within the range expected for histone–DNA interactions. The classical B-form DNA is a linear double helical structure (Fig. 1.4(a)). Thus DNA coiled tightly around the histones in the core particle cannot be precisely of this form. The DNA molecule, however, has sufficient flexibility to be bent and deformed to the radius of the DNA coil around the core particle. In the low resolution models described above for the structure of the core particles DNA is thought to be in modified B-type structure uniformly deformed to coil at a radius of 4.5 nm.

2.5 STRUCTURAL ROLES OF DIFFERENT HISTONES

It was initially thought that the four histones H2A, H2B, H3 and H4 were equally important in generating the structure of the nucleosome or core particle. Results from several laboratories, however, now show that many of the structural features of the nucleosome and chromatin can be generated by histones H3 and H4 alone. Complexes of higher molecular weight DNA with H3 and H4 give a 145 bp DNA particle on digestion with staphlococcal nuclease, very similar patterns of subnucleosome DNA lengths with increasing time of digestion and a 10.4 base single stranded DNA pattern on digestion with DNase I. Complexes of histones H3 and H4 with DNA give an almost identical fibre X-ray diffraction pattern as found for chromatin fibres. H3 and H4 appear, therefore, to be the major structural proteins of the nucleosome and determine the 145 bp lengths of DNA protected in the core particle. These findings are consistent with the rigid sequence conservation found for histones H3 and H4 although it is not understood how histones H3 and H4 are able to control an amount of DNA of substantially larger mass. The situation is quite different in viruses, for example, where the coat proteins, which determine the structure of the virus, are present in substantially larger amounts than the nucleic acid component, RNA or DNA. This central structural role for histones H3 and H4 raises questions concerning the roles of histones H2A and H2B in chromatin structure. It has been suggested that H2A and H2B add further stability to the nucleosome structure and, in effect, complete the nucleosome structure. It is equally probable that the core particle structure formed by DNA and H3 and H4 provide the binding sites for H2A and H2B which are then involved in functions outside the core particle structure. Such functions may involve the linker DNA or the DNA associated with higher order chromatin structures.

2.5.1   Interactions of Histones in Core Particles

At present we have insufficient data from neutron scatter or X-ray diffraction studies to work out the arrangement of histones in the core particle. It is expected, however, that X-ray and neutron diffraction studies of core particle crystals will eventually give a high resolution model. Chemical cross-linking of histones in chromatin should give information on contact between histones but only if short cross-links are used. Although dimethyl suberimidate has been used to demonstrate the existence of specific histone complexes, this molecule is too long to give data on close contacts between histones. Using ultraviolet light and tetranitromethane as cross-linking agents, close points of contact have been identified between histones in the pairs (H2A, H4) and (H2A, H2B). Cysteine bridge formation also shows the presence of H3–H3 contacts between the apolar regions of these molecules.

From the histone sequences discussed in Chapter 1 it has been suggested that the basic N-terminal regions of the core histone H2A, H2B, H3 and H4 are major sites of interaction with DNA. Several models for nucleosomes have incorporated these ideas and show all of the basic N-terminal regions of the eight core histones hooked around the DNA. Such models are now thought to be too simple and do not provide an explanation as to why the basic N-terminal regions of H2A and H2B are variable while those of H3 and H4 are rigidly conserved. On trypsin digestion of chromatin it is found that the N-terminal regions of the core histones are preferentially digested which is consistent with them being on the outside of the nucleosome and accessible to the enzyme. Digestion of these N-terminal regions, however, does not result in an uncoiling of the DNA around the nucleosome to the linear form. Also core particle-like structures have been reformed from 140 bp DNA and the core histones form which the N-terminal regions have been cleaved. This suggests that interactions of the apolar central and C-terminal regions of the histones are important in stabilizing the nucleosome structure. It is also known from spectroscopic studies that the basic regions of the histones H2A and H2B are not bound in the core particle. It would appear from these observations that the conserved apolar central and C-terminal regions of H2A, H2B, H3 and H4 are major sites of interaction both between histones and between histones and DNA. The apolar regions of these histones contain 70 % of the arginine residues which are strongly binding to DNA. The N-terminal regions of the histones contain a large proportion of lysine residues which are more weakly binding to DNA.

2.6 SUMMARY OF CORE PARTICLE AND NUCLEOSOME STRUCTURE

From the properties and interactions of histones it would appear that the highly conserved histones H3 and H4 interact with 146 bp of DNA to give a structural unit which is completed by the binding of the conserved apolar regions of the histones H2A and H2B to give the conserved core particle. The sequence conservation of the central apolar region of H1 suggests that this region is also involved in interactions within the nucleosome. In the staphylococcal nuclease digestion process H1 is lost from the nucleosome when the DNA from 168 to 146 bp is digested away. It has been suggested from electron microscopic studies that H1 is located on one side of the nucleosome and 'seals off' two turns of DNA by binding across the edge of the disc shaped particle. This proposal is consistent with the properties of H1 if the apolar central region of H1 binds across the edge of the particle.

An outline of the structure of the nucleosome is given in Fig. 2.6. This shows two turns of DNA with 84 bp of DNA per turn and involves a histone core of the tetramer (H3, H4)$_2$ and the apolar regions of H2A and H2B. H1 is bound to the edge of the disc across the two turns of DNA and completes the 168 bp particle. The variable N-terminal regions of H2A and H2B, and the variable N- and C-terminal regions of H1 are thought to act outside of this conserved structural unit.

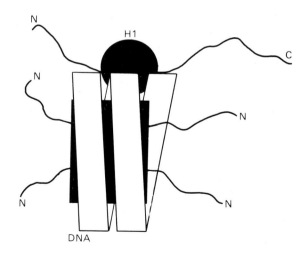

Fig. 2.6 Outline of structure proposed for the nucleosone. Two turns of DNA with 84 bp per turn are sealed off by H1. The path of the linker DNA joining adjacent nucleosomes is not known yet.

2.7 HIGHER ORDER STRUCTURE

Many electron microscope studies of chromatin structure have consistently described filaments with diameters of about 10 and 30 nm as well as even higher order structures. The 10 nm filament, called the nucleofilament, consists of a linear array of nucleosomes and is observed in solution under conditions of low ionic strength. With increase in ionic strength, either monovalent or divalent cations, the nucleofilament undergoes a transition to the '30 nm' filament. Many electron microscopic studies have reported the widespread occurrence of the '30 nm' filament in interphase and metaphase nuclei. This coil filament with a diameter of 33 nm is shown (Fig. 2.7) for mouse liver chromatin.

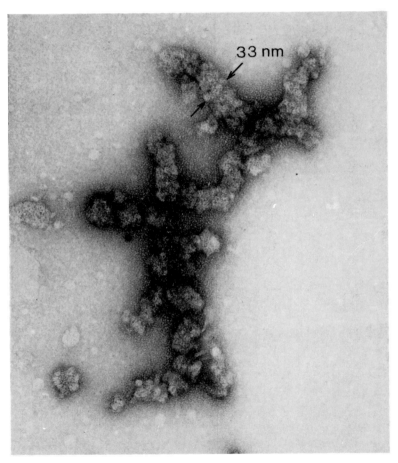

Fig. 2.7 Electron microscope picture of a fragment of mouse liver chromatin showing the contracted supercoil. (Courtesy Miller F., Igo-Kemenes T. & Zachau H. G.)

2.7.1 Arrangement of Nucleosomes in the Nucleofilament

The finding that the chromatin core particle is a disc $11 \times 11 \times 5.5$ nm has raised questions as to how the discs are arranged with respect to each other in the nucleofilament. X-ray and neutron scatter of the nucleofilament in solution at low ionic strength, less that 20 nm NaCl, gives a mass per unit length equivalent to 1 nucleosome per 10–11 nm. This corresponds to a DNA packing ratio of about 7:1. The core particle arrangement which agrees best with the neutron scatter data is with the discs close to edge-to-edge is shown in Fig. 2.8(a).

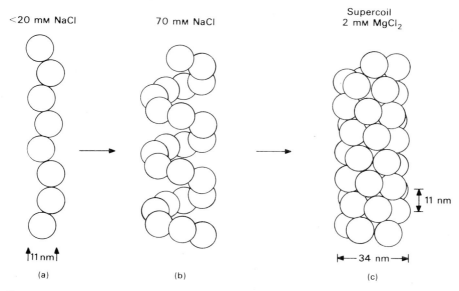

Fig. 2.8 Ionic strength induced transition from the extended form of chromatin at low ionic strength to the most compact supercoil of the nucleosome at high salt. In (a) the discs are edge to edge; in (b) and (c) the arrangements of the discs are not known and the circles simply represent nucleosomes.

2.7.2 '30 nm' Chromatin Filament

The ionic strength induced transition from the '10 nm' nucleofilament has been studied by neutron scatter techniques and by electron microscopy. From the neutron studies it was found that above 20 mm NaCl the nucleofilament made an abrupt transition to a higher order structure with a transverse radius of gyration of about 9.0 nm and a packing of 2 nucleosomes per 10 nm. With increasing ionic strength the radius of gyration increased slightly to 9.5 nm while the packing ratio increased to 6–7 nucleosomes per 10 nm. This behaviour suggests a family of supercoils of the nucleofilament that becomes more tightly coiled with increasing ionic strength. The diameter of the hydrated state of the most compact supercoil was 34 nm which corresponds to the range of diameters of 25–33 nm found for the dehydrated form of the supercoil in electron micrographs. This salt-induced transition is indicated schematically in Fig. 2.8. Electron microscopy studies have led to similar proposals for the initial (Fig. 2.8(a)) and final (Fig. 2.8(c)) states of this transition but suggest a different

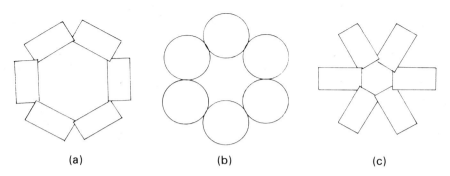

Fig. 2.9   Possible arrangement of nucleosomes in the 34 nm diameter supercoil.
(a) nucleosome discs facing out from the helix axis; (b) nucleosome discs facing along
the helix axis, and (c) nucleosome discs arranged radially.

mode of compaction to that given for the intermediate supercoil of Fig. 2.8(b).
In this alternative process of compaction the nucleosomes in the 10 nm nucleofilament undergo a transition to a regular zigzag arrangement of edge-to-edge
nucleosomes. Further compaction involves a progressive rearrangement of the
zigzag of nucleosomes to the most compact form of the supercoil. This most
compact form, Fig. 2.8(c), corresponds to the structure proposed for chromatin
from neutron fibre diffraction, X-ray and electron-microscope studies which
attribute the well-known semi-meridional arc at 10–11 nm to the 11 nm pitch
of the supercoil. Because the nucleosomes are disc-shaped (Fig. 2.6) they can
be arranged in many ways in the 34 nm supercoil. In Fig. 2.9 three orthogonal
arrangements of the discs, as seen looking down the axis of the supercoil, are
shown. In (a) the discs are facing out from the surface of the supercoil, in (b)
the faces are pointing along the axis of the supercoil while in (c) they are arranged radially. The evidence so far suggests that the first arrangement is
unlikely to be correct. A possible arrangement is that given in (c) though an
intermediate arrangement between (b) and (c) with the discs inclined cannot be
excluded. Further studies of this salt-induced transition are required to determine the precise arrangement of the disc-shaped nucleosomes in the supercoil,
the interactions of histone H1 and the core histones which generate the supercoil
and the location of histone H1.

2.7.3   Histone H1 and Higher Order Chromatin Structures

There is much evidence to show that the H1 histone is involved in higher order
chromatin structures. The compaction of chromatin under a variety of conditions has been correlated with the presence of H1. The form of H1 (Fig. 1.15)
of three well-defined domains suggests that H1 is multi-functional. It has been

suggested that the H1 molecule seals off two turns of the DNA coil around the histone octamer of the conserved core particle structure (Fig. 2.6) and this function probably involves the conserved central apolar region of H1. Although the 34 nm supercoil can be generated by interactions of histone H2A, H2B, H3 and H4 this second order structure is stabilized by additional interactions of H1. Further chromosomal coilings are thought to involve further interactions of the H1 molecule probably through the variable basic N- and C-terminal regions of H1 which are also subject to the reversible phosophorylations of serines and threonines.

2.8 STRUCTURE OF THE METAPHASE CHROMOSOME

For many years it had been thought that the metaphase chromosome was made up of a linear series of coiled coils, i.e. the long DNA molecule was packaged into a first order coil which was then itself coiled etc. until the DNA molecule was contained within the length of the metaphase chromosome. Recent evidence questions whether this simple concept for longitudinal packaging of DNA is correct. There have been several studies which suggest that the DNA in eukaryotic chromosomes is in the form of loops. Gently isolated metaphase chromosomes have been centrifuged through a cushion of dextran sulphate, which dissociates the histones, on to an electron microscope grid on the bottom of the centifuge tube. An electron micrograph of the resulting histone depleted metaphase chromosomes is shown in Fig. 2.10. This appears to show a matrix of non-histone protein with the same shape as the metaphase chromosome. The DNA radiates from this matrix of protein and forms a holo of free DNA. At higher magnification the DNA could be seen to form loops which enter and leave the protein matrix at the same point. The histone free DNA loops are in the range of 10–30 μm in length. Other evidence for the existence of long range constraints applied to chromatin has come from the dissection of chromatin with staphylococcal nuclease and restriction enzymes. Analysis of the sizes of the chromatin pieces released, led to the proposals of the domain theory whereby the chromatin is constrained to form domains of average size containing 35 000–70 000 bp DNA, i.e. 12–25 μm. It is, of course, of considerable interest to know whether there is just one gene per domain. Both electron microscope and biochemical studies have led to very similar conclusions concerning long range order in chromosomes. These conclusions may establish a link between the behaviour of individual chromosomes in cells and the visible array of many copies of chromosomes arranged side-by-side in the giant polytene chromosomes of the salivary glands of *Drosophila* and other insects in which active genes are contained in chromatin domains or loops are recognized and their gene or genes activated individually (see Chapter 6).

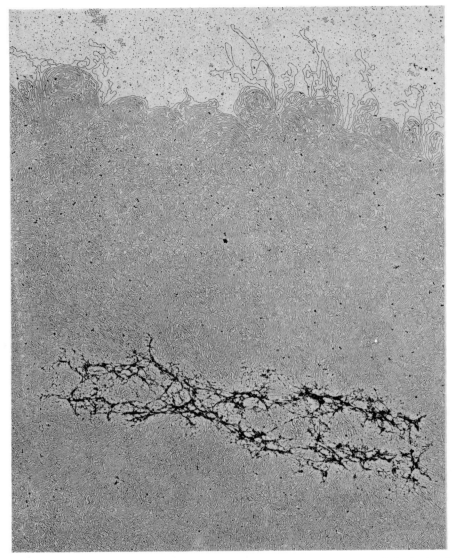

Fig. 2.10 Electron microscope picture of a He la metaphase chromosome fully depleted of histones. 10 to 30 μm length DNA loops extend out from a matrix of non-histone chromosomal protein. (From Paulson J. H. & Laemmli U. K. (1977) *Cell* **12**, 817–828.)

2.9 CHROMOSOME BANDING

During the last decade an important new development in the study of chromosomes has been the discovery of staining techniques which are able to resolve a banded pattern on mitotic chromosomes from many organisms. Since this pattern varies from one chromosome to another within a set, but is similar for any two homologous chromosomes, these techniques have permitted chromosome identification with a precision never previously possible. In earlier work with orcein or Feulgen stains, only 6 of the 23 pairs of human chromosomes could be unambiguously identified, whereas today it is relatively easy for an experienced cytogeneticist to identify each human chromosome with certainty. Although unquestionably the greatest contribution of chromosome banding techniques has been for the purposes of chromosome identification and diagnostic cytogenetics, the molecular basis for the procedure is of some interest, particularly as regards its possible relationship to fundamental aspects of chromosome structure and function.

We should make clear at the outset that these bands must not be confused with periodicities of chromatin structure imposed by nucleosomes, or with the chromomeres and bands visible in many polytene and meiotic chromosomes, especially in the giant chromosomes of many insect tissues. The bands which we are here considering most frequently number no more than about 20 per chromosome (see Fig. 2.11) and are clearly related to variations in chromatin organization at an extremely gross level on or in the mitotic chromosome. (Although it is interesting to remark that Q and G banding of prophase chromosomes can display numerous bands which may be analagous to chromomeres). Since the nomenclature of the various banding methods is slightly complex and the various techniques yield somewhat different patterns in the same chromosomes, we will now list the methods and discuss each in turn in its relation to chromosome structure (greater detail will be found in the review of Evans (1977)).

Five separate banding patterns can presently be revealed in vertebrate mitotic chromosomes and these are variously termed C bands, Q bands, G bands, R bands and N bands. Some of these bands are common to two or more techniques, others are apparent only following one particular staining method.

2.9.1 C Bands

These bands are so termed because they involve the centromere regions of chromosomes consisting of constitutive heterochromatin, although such bands are also present in chromosome arms, as in the human Y chromosome and in some autosomes of other non-human vertebrates. C bands are revealed by DNA denaturation, and strong alkali or formamide are common reagents employed to accomplish this. What seems to happen on exposure of chromosomes to such agents is that most of the chromatin becomes decondensed, while the C band material remains tightly condensed. It is now known that C band chromatin contains a special type of DNA composed of very highly repetitious short sequences (see Figs. 2.12, 2.13).

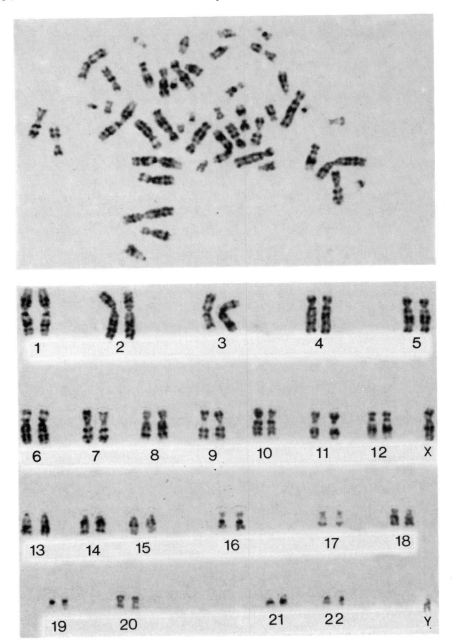

Fig. 2.11 Giemsa G-band staining of human chromosome complement (male). Notice that both members of a pair show identical banding patterns, but marked differences exist between pairs, permitting all pairs to be easily distinguished. Kindly provided by Prof. H. J. Evans. (From Evans H. J. (1973) *Br. med. Bull.* **29**, 196.)

2.9.2 Q Bands

Both Q bands and G bands are broadly representative of the same material and the patterns, although differently produced, are more or less the same. As we shall have cause to discuss later, R banding is actually the reverse of this pattern. Bands of the Q type are ones which show bright fluorescence with a group of dyes known as quinacrines (thus the name Q bands). Differential staining and fluorescence of chromatin with quinacrine can be observed without heating or denaturing of the material, and although small parts of chromosomes seem to fluoresce with the dye because they contain DNA rich in A–T bases, in the main the bands seem to represent simply altered states of chromatin compaction.

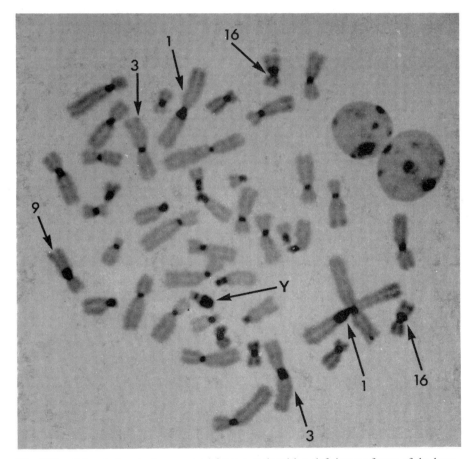

Fig. 2.12 C-banded metaphase spread from a male with a deficiency of part of the long arm of chromosome 9, and anomalies in some other chromosomes. Chromosomes with major C-bands, including the Y chromosome, are arrowed. Kindly provided by Prof. H. J. Evans. (From Evans H. J. (1973) *Br. med. Bull.* **29**, 196.)

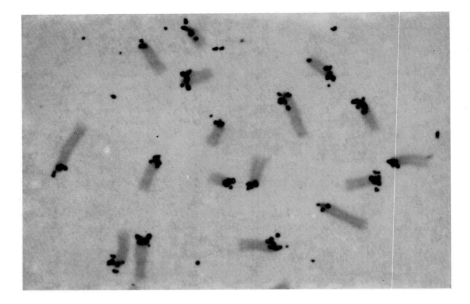

Fig. 2.13   In-situ hybridization of labelled mouse satellite DNA on mitotic cells. Notice that the label is concentrated at the centromeres of the chromosomes. Kindly provided by Prof. J. G. Gall. (From Pardue M. L. & Gall J. G. (1971) *Chromosomes Today* **3**, 47.)

2.9.3   G Bands

Staining of chromosomes with Giemsa will not normally produce a banded pattern, but if the chromosomes have been pretreated with trypsin, pronase, warm saline, heat, or alkali solutions, a pattern of G bands can be resolved with Giemsa staining, (Fig. 2.11). The staining pattern is essentially similar to that apparent with quinacrine fluorescence, and over 50% of the chromatin of the human chromosome set will stain in this way. No clear understanding of the mechanism of G band staining has as yet emerged, but it seems to be related to the tightness of the condensation of the chromatin.

2.9.4   R Bands

R bands are so called because the pattern of staining is the reverse of that induced by the Giemsa and Quinacrine methods. R bands can be produced by fixing chromosomes in conventional ways, followed by incubation at 87–89°C for 10 min. in saline solution. Following this pretreatment R bands are apparent following Giesma staining.

2.9.5   N Bands

N bands are bands which can be induced in the nucleolar organizing regions of chromosomes, that is the groups of genes coding for ribosomal RNA. One or two different methods have been successfully used to display N banding, one being Giemsa staining after vigorous acidic digestion of chromatin. A more interesting approach has been the use of ammoniacal silver staining to produce such bands on active DNA, and it is useful to note that 'in-situ' hybridization with labelled 18S and 28S RNA produces a similar location map for the ribosomal cistron clusters.

The only useful general comment about Q, G and R banding is to say that the banding pattern of a particular chromosome in related species is similar, and seems to be stable over long periods of evolution. By contrast, C banding patterns, especially in the non-centromeric regions are very variable in different species.

2.10 HISTONE MODIFICATIONS AND CHROMATIN STRUCTURE

As shown in the previous chapter the sites of reversible acetylations of the core particle histones, H2A, H2B, H3 and H4 are all located in their basic N-terminal regions and the reversible growth associated sites of phosphorylations of the very lysine-rich histone H1 are in the basic N- and C-terminal regions of this molecule. These reversible chemical modifications have been correlated with important functions and with structural changes of chromosomes; acetylation with DNA replication and with transcription and H1 phosphorylation with chromosome condensation and mitosis (Chapter 5).

Until recently it was very difficult to obtain useful quantities of the most highly acetylated forms of histones for structural studies because they occurred in the overall histone populations in very small amounts. This situation has now been changed by the findings by Ingram's laboratory that the addition of *n*-butyrate to cultured cells resulted in an accumulation of the more highly acetylated histones and to the arrest of cell growth. It has been shown that the effect of butyrate is to inactivate histone de-acetylases and this inactivation leads to an accumulation of the acetylated histones e.g. adding butyrate to make the solution 7 mM to the cells in the log phase of growth for 24 h gave approximately equimolar amounts of the non, mono, di, tri and tetra acetylated forms of H4. So far the 'acetylated' chromatin has been used for two studies; firstly, it has been shown that DNase I (see sections 2.4.1) preferentially digests regions of chromatin containing the tri and tetra acetylated histone H4. This adds further weight to the correlation between selective DNase I digestion of active chromatin and the presence of acetylated histones. Secondly, initial studies have been made of the effect of this level of acetylation on the structure of the nucleosomes and core particles. So far no significant structural differences have been observed between these 'acetylated' nucleosomes and control nucleosomes. It may be that the full structural effect of acetylation on the nucleosome or core particles requires that all the histones are fully acetylated. A more probable explanation is that histone acetylation does not effect the structure of the core particle, though it may reduce its stability, but exerts its effect at the level of the 34 nm supercoil by modulating interactions of the N-terminal regions of histones with the linker DNA between core particles. This would be consistent with the conclusions (section 2.5.1) that interactions of the central and C-terminal regions of histones H2A, H2B, H3 and H4 stabilize the structure of the core particle.

As will be discussed in Chapter 5, reversible phosphorylation of H1 has been correlated with chromosome condensation though the precise mechanisms whereby the condensation is achieved are not known at present i.e. whether H1 acts as a cross-linking molecule, which would be consistent with its structure and properties, on whether H1 phosphorylation releases constraints applied by the H1 molecule in order to decondense chromatin.

A scheme which encompasses our present understanding of chromatin structure, reversible chemical modifications, HMG proteins and metaphase chromosome scaffold proteins is given in Fig. 2.14. In this scheme the basic form of inactive chromatin is given as the 34 nm supercoil. The transition of this supercoil to the 'active' chromatin structure for DNA processing involves acetylation of histone H2A, H2B, H3 and H4, the addition of HMG proteins 14 and possibly 17. H1 is probably dissociated in this process. The HMG proteins may bind to the DNA binding sites exposed by the acetylation of the basic N-terminal regions of these histones and the loss of H1. As will be explained in Chapter 5 it is not known whether active chromatin contains nucleosome-like structures or is fully extended. Above the 34 nm supercoil the process of chromosome condensation involves H1 phosphorylation and for the fully compacted metaphase chromosomes the scaffold proteins.

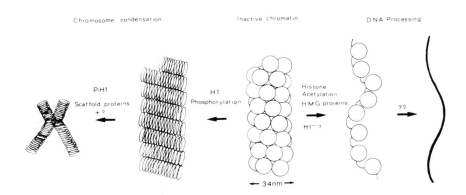

Fig. 2.14 Outline scheme proposed for the effects of histone modifications on chromosome structure.

GENERAL REVIEWS

MORE ADVANCED READINGS

Allfrey V. G. (1977) Post-synthetic modifications of histone structures. In *Chromatin and Chromosome Structure* (eds Li H. J. & Eckhardt R.) pp. 167–91. Academic Press, New York.

Cold Spring Harbor Symp. (1977). Chromatin, **42**, (1 & 2).

Dixon G. H., Candido E. P. M., Honda B. M., Loui A. J., Macleod A. R. & Sung M. T. (1974) The biological roles of post-synthetic modification of basic nuclear proteins. *CIBA Found. Symp.: The Structure and Function of Chromatin*, Vol. 28, pp. 229–58. Elsevier, Amsterdam.

Evans H. J. (1977) Some facts and fancies relating to chromosome structure in man. *Adv. Human Genet.* **8**, 347–438.

Felsenfeld G. (1978) Chromatin *Nature* **271**, 115–22.

Finch J. T. & Klug A. (1976) Solenoidal model for superstructure in chromatin. *P.N.A.S.* **73**, 1897.

Finch J. T., Lutter L. C., Rhodes D., Brown R. S., Rushton B., Levitt M. & Klug A. (1977) Structure of nucleosome core particles of chromatin. *Nature* **269**, 29–36.

Goodwin G. M., Walker J. M. & Johns, E. W. (1978) The high mobility group (HMG) nonhistone chromosomal proteins. In *The Cell Nucleus* (ed. Busch H.) pp. 181–219. Academic Press, New York.

Kornberg R. D. & Klug A. (1981) The nucleosome. *Scientific Amer.* **Feb. 1981**, 48–60.

Lutter L. C. (1978) Kinetic analysis of Deoxyribonuclease I cleavages in the nucleosome core: evidence for a DNA super helix. *J. Mol. Biol.* **124**, 391–420.

Lutter L. C. (1979) Precise location of DNase I cutting sites in the nucleosome core determined by high resolution gel electrophoresis. *Nucl. Acids Res.* **6**, 41–56.

Matthews H. R. (1977) Phosphorylation of H1 and chromosome condensation. In *The Organization and Expression of the Eukaryotic Genome* (eds Bradbury E. M. & Javaherian K.) pp. 67–80. Academic Press, London.

Paulson J. R. & Laemmli U. K. (1977) The structure of histone-depleted metaphase chromosomes. *Cell* **12**, 817–28.

Suau P., Bradbury E. M. & Baldwin J. P. (1979) Higher order structures of chromatin in solution. *Eur. J. Biochem.* **97**, 593–602.

Suau P., Kneale G. G., Braddock G. W., Baldwin J. P. & Bradbury E. M. (1977) A low resolution model for the chromatin core particle by neutron scattering. *Nucl. Acids Res.* **4**, 3769–86.

Von Holt C., Strickland W. N., Brand W. F. & Strickland M. S. (1979) More histone sequences. *Review F.E.B.S. Lett.* **100**, 201–18.

Chapter 3
The Genomes of Bacteria, Viruses, Plasmids, Mitochondria and Chloroplasts

This book is chiefly about the DNA and chromosomes of eukaryotic nuclei, but a chapter must be included on the other interesting forms of genetic material found in nature. These include the genomes of the cytoplasmic organelles of eukaryotes, that is of mitochondria and chloroplasts, bacterial and viral genomes, and the groups of genes which make up the genetic factors found commonly in bacteria, the plasmids. There are many fundamental ways in which all these forms of DNA (and, in a few cases, RNA) differ from the genetic material found in eykaryotic nuclei. The first is that these genomes are not packaged, strictly speaking, in the form of chromatin, although they are often referred to as chromosomes. With the special exception of some animal viruses such as SV40, the genomes of plastids, viruses and prokaryotes are less obviously conjugated with basic protein to form nucleosomes, although in the laboratory of Francois Gros in Paris, and elsewhere, a histone-like protein, termed HU protein, has been found to complex with DNA in bacteria. The resultant conjugate seems to form a beaded structure but appears to be less stable than eukaryotic chromatin, however. Secondly, there being no nuclear membranes intervening between genome and ribosome, the acts of transcription and translation of the genetic code are much more closely linked. Ribosomes are often involved in translating the messenger RNA before it has been completely transcribed or physically freed from the DNA. This also implies that there is less clearly an intermediate class of large Hn RNA acting as precursor to the messenger RNA.

Thirdly, all of the genomes which we will discuss in this chapter are very small by eukaryotic standards (see Table 3.1) and great economy in the use of the DNA is often apparent. This contrasts with the rather lax way in which the often massive genomes of higher cells are organized and utilized as functional genetic message. And so, especially amongst the smaller viruses, we encounter genes overlapping along the same length of DNA, and also, in many cases, both strands of the same section of DNA being utilized as sense coding sequence. The small size of these genomes and the lack of cellular differentiation also renders many of the genetic control systems presumed to occur in eukaryotes unnecessary for bacteria and viruses. The operons and regulons of bacteria are almost certainly simpler than the eukaryotic genetic control mechanisms which still tantalize geneticists.

A final point to make is that the small size of these genomes and the ease of culture of many bacteria and viruses, has rendered their DNA sequences more amenable to study. At the time of writing more than one viral genome has been entirely sequenced, as have large tracts of *E. coli* genome.

Table 3.1   Physical characteristics of prokaryotic and other genomes.

Organism or factor	Mol Wt of genome	Shape of genome	Length of genome μm	No. of genes in genome approx.
Mitochondrion		circle	5–30	70?
Chloroplast		circle	50	100s?
E. coli	2×10^9	circle	1100	1800
T4 phage	11×10^7	linear	60	150
Lambda phage	3.2×10^7	circle/linear	16	35
SV40 virus	3.4×10^6	circle	3.4	10?
S.S. DNA phage (ϕX174)	1.7×10^6	circle	1.7	8
RNA phage (f2)	1.3×10^6	linear	1.0	3
Sex factor plasmid	4×10^7	circle	20	12–100?
R factor plasmid	$10–70 \times 10^6$	circle	5–30	

3.1   BACTERIAL GENOMES

The bacterium which is best known genetically is, of course, the enteric bacillus *Escherichia coli*, and strain K12 has been very intensively studied. Much of the evidence which will be quoted is therefore derived from experimental work on this amenable and versatile organism.

3.1.1   Structure of the genome

The *E. coli* genome is a closed circle of double stranded DNA. This DNA is not enclosed within a nucleus but must obviously be quite tightly folded to fit within the cell. Indeed there is good evidence that the DNA is highly folded, occupying only part of the cell. It is sometimes referred to as the nucleoid. Immediately after division a bacterial cell may possess only one copy of the genome and therefore one copy of each gene. But this statement can be confusing, since, in a rapidly growing culture of bacteria, DNA replication may anticipate division and a single cell may possess two circles of DNA each already part way through a further round of replication. So although a bacterial cell may at times have only one copy of the genome, many or even most cells possess more than one copy of each gene at any one moment in time. About 1100 μm long and comprising some 3.4×10^6 pairs of nucleotides, there is space in the *E. coli* genome for a few thousand separate structural genes, but present information suggests that the total number of genes may be nearer to 1800. About 650 loci have been allotted map positions in the circular map of the K12 genome (see Fig. 3.1). A study of this diagram will also reveal an interesting and curious fact, namely that genes are not distributed randomly on the genome. In fact the loci presently mapped are, in many cases, highly clustered, with the result that quite long tracts of the DNA are left empty and apparently silent. Compare, for example, the clustering of some 50 loci between 80.5 and 85.5 minutes, with the detection of only 2 loci on the equivalent length of 30–35 minutes on the K12 genome. It seems likely that this uneven distribution of mapped genes is not a quirk of the genes so far amenable to mapping, but reflects real differences in the utilization of the bacterial genome. Just as very large amounts of the eukaryotic genome seem to be void of structural genes, and indeed large parts of each section of chrommomere DNA (see discussion of this topic in Chapter 7), so it may also be with the bacterial genome. As suggested for the former situation, these lengths of silent DNA may have a special role in the folding and packing of the bacterial genome into the fairly condensed space occupied by the nucleoid.

3.1.2 Mechanisms of Genetic Variation in Bacteria

Bacteria do not display as part of their life cycles, the elaborate sexual methods designed to ensure production of genetic variation in higher animals and plants. True they possess a sort of sexuality in terms of mating strain, but, as we shall shortly emphasize, the mating process is in no way analogous to that found higher up on the evolutionary scale. Prokaryotes do experience considerable genetic variation, however, and we will now briefly discuss the four distinct ways in which this may come about.

MUTATION

Along with all living organisms, bacteria are susceptible to mutation. If a culture of *E. coli*, sensitive to a particular virulent phage, is plated out on agar in the presence of that phage, most of the cells will be lysed. But some colonies will grow up, and these will prove to be resistant to the phage. It was the classical *fluctuation test* of Luria and Delbruck which demonstrated that mutation in this and similar situations is spontaneous and is not itself induced by exposure to the phage or any other selective force. Mutation of the DNA is therefore occurring constantly at a low rate. We will assume here that our reader is aware of the fact that mutation involves damage to the DNA code, either by deletion or substitution of one or more nucleotides, and is induced by many factors, chemical and physical, of which cosmic rays in Nature and ultraviolet light in research are the best known. All we need emphasize here is that the spontaneous mutation rate for any one gene in bacteria is about 2×10^{-8} mutations per bacterium per generation. Assuming the total gene number is close to 2000, the number of expected (though not necessarily detectable) mutations per generation is about one per half million bacteria.

We should say a little more about the mechanics of mutation, and in particular discuss the role of the DNA repair enzymes. Repair of DNA after mutation is itself a highly complex affair, and may involve a wide spectrum of different molecules. But at its simplest it involves the action of a nuclease enzyme

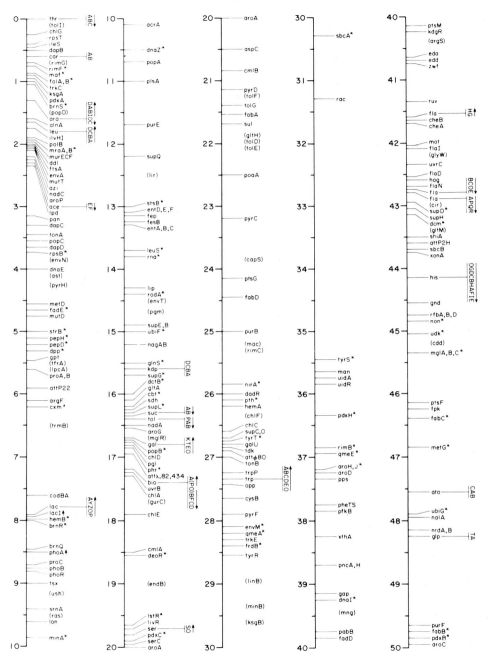

Fig. 3.1 Map of the chromosomes of *E. coli* K12 shown in linear form. (From Lewin B. (1977) *Gene Expressions Vol.* **3**. John Wiley & Sons, Chichester.)

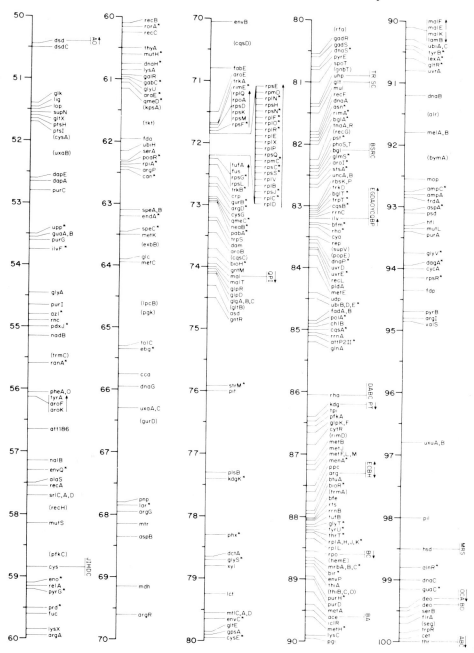

which specifically recognizes the DNA error and cuts out the intruding base, followed by the operation of a repair polymerase enzyme which inserts the correct base in the vacant space in the chain. Different strains and species of bacteria differ greatly in the efficiency of this repair process, some, the highly inefficient ones, being therefore extremely sensitive to u.v. light, while others, with very effective repair enzymes, are very insensitive. Indeed strains of bacteria are known which can withstand levels of ionizing radiation highly lethal to man, not because they harbour lead shielding in their cell walls but because their DNA repair systems are extremely effective! The really important point to grasp about DNA repair is that it takes much of the unpredictability out of the mutation phenomenon. The mutation rate itself may be adjusted by modulation of effective repair. So the basis of natural selection and evolution, the mutational event, must now be seen as something under the control of the cell and itself open to variation in rate through natural selection.

TRANSFORMATION

One of the most renowned experiments in the history of science was carried out by Griffiths in 1928. While studying the pathology of pneumococci in mice, he found that both virulent and non-virulent strains of these bacteria could be isolated. Virulence could also be eliminated by heating the bacterial culture. But to his surprise, after mixing together living non-virulent pneumococci and heat-killed virulent pneumococci, mice injected with this mixture subsequently died. Moreover *living virulent* bacteria could be isolated from the dead mice. It seemed that, in some way, the virulent character had been transmitted from the dead bacteria to living ones, resulting in live virulent pneumococci. Following up Griffith's important experiment, Avery, Macleod and McCarty were able to demonstrate in 1944 that the material which carried the character from one bacterium to another was DNA, identifying this molecule as the genetic material. Their experiments also showed, as had those of Griffiths, that DNA was capable of acting as a 'transforming principle' in bacteria, moving from one cell to another and, at least in part, being permanently accepted into the genome of the recipient bacterium. This, then, is bacterial transformation, the transmission from cell to cell of DNA molecules and the induction of permanent genetic change in the recipient bacterium by that DNA.

It is probable that, in nature, transformation is a rare event, and even in the test tube its incidence is low. Incubation of a culture of bacteria in a medium containing suitably prepared DNA continues to yield only a few transformed cells. The precise mechanism of uptake of the DNA into the cell is not known, nor is it clear whether single or double strands are accepted. What is clear is that, following passage of a sequence of DNA into a bacterial cell, the new DNA may then be involved in recombination with the host genome. Such recombination may be either additive or in the nature of an exchange, so that either there is a total gain of DNA, or some host cell genes may be lost in the exchange. DNA recombination by exchange is normally referred to as *substitutive* recombination. The process of additive recombination is believed to occur on the lines of a model proposed by A. M. Campbell, actually suggested for the integration of phage lambda into the bacterial genome. It is discussed in our section on phage lambda and illustrated in Fig. 3.18.

Transformation is therefore a mechanism whereby a bacterial cell can acquire new genetic properties by the acquisition of DNA molecules from the surrounding medium. One point of the greatest significance for 'genetic engineering' is that such DNA need not necessarily be bacterial in origin, although most experiments which fall into this classification now utilize bacterial plasmids as carriers.

CONJUGATION

Bacterial conjugation is the movement of DNA from a donor to a recipient cell by a process which demands physical connection between the two cells. Although it is often described as a mating process, it is a phenomenon which depends entirely on the possession and action of a plasmid within the donor bacteria. So we might justifiably describe bacterial sexuality as a contageous infection. Details of the process are discussed in this chapter under the heading of the sex factor plasmid itself, and so we will here only emphasize the aspects which have a special bearing on genetic variation.

Conjugation, then, is, at its simplest, the transference from one cell to another of a plasmid F, the sex factor. It requires the formation of an F pilus, a structure coded by the plasmid itself. But in some strains of bacteria the sex factor is integrated in the host genome. When such HFr strains indulge in conjugation, part of the bacterial genome itself is transferred to the recipient F^- cell, and only very rarely the sex factor. So this aspect of conjugation is somewhat analogous to transformation and the bacterial genes transferred are available for recombination with the genome of the recipient cell just as we discussed in transformation.

But yet another variant of conjugation involves F prime factors (F'), which, though not integrated themselves, do carry some bacterial genes within or attached to the plasmid sequence. Conjugation involving F' bacteria with F^- strains is characterized by efficient transfer of both sex factor and some bacterial genes. Some of these F' plasmids are themselves defective since they have lost portions of their own plasmid DNA, presumably in exchange for bacterial genes. Others are however, intact, and may be quite large, carrying long lengths of bacterial DNA, presumably due to faulty excision of the plasmid from the host cell genome.

With both HFr and F' bacteria, bacterial genes are transferred to recipient cells and may become a permanent part of the genetic complement of that cell by recombination. Moreover, just as HFr transfer of bacterial DNA is analagous to simple transformation, so the F prime transfer is analogous to our next topic, transduction, since in each case a plasmid is acting as a carrier system for bacterial genes.

TRANSDUCTION

Transduction is the conduction of bacterial genes from one cell to another by a carrier phage. It was discovered in 1951 by Zinder and Kederberg during studies of recombination in *Salmonella*. Using a Davis U-tube, which, as illustrated in Fig. 3.2, consists simply of a hollow glass U-shaped tube with an

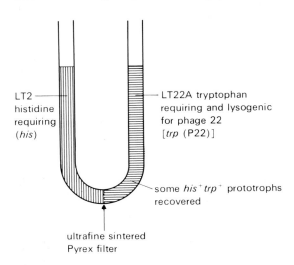

LT2
histidine
requiring
(*his*)

LT22A tryptophan
requiring and lysogenic
for phage 22
[*trp* (P22)]

some *his⁺ trp⁺* prototrophs
recovered

ultrafine sintered
Pyrex filter

Fig. 3.2 The Davis U-tube with which the process of transduction was first demonstrated. (After Smith-Keary P. F. (1975) *Genetic structure and Function*. Macmillan, London.)

ultrafine filter inserted in its centre, they placed a separate strain of *Salmonella* in each arm of the tube, one strain requiring histidine for growth, the other requiring tryptophan. The filter in the U-tube was sufficiently fine to prevent bacterial cells passing between the two sides and also prevented physical contact between cells of the two strains. After the liquid in the tube had been sucked back and forth a few times the bacterial strains were recovered. To the astonishment of the experimenters, they were then able to recover from amongst the tryptophan-requiring strain many cells which would now grow without tryptophan. Since the tryptophan requiring cells carried phage 22 as a prophage, and the histidine requiring cells were lysed by this phage, Zinder and Lederberg rightly argued that a few phage must have been released by the tryptophan-requiring *Salmonella*, the phage then passing through the filter and infecting the sensitive histidine-requiring strain. Following lysis, the phage were then presumed to have picked up pieces of bacterial DNA when they packaged their own genomes within new capsids, releasing these genes when they re-entered some of the tryptophan-requiring *Salmonella*.

This phenomenon is now termed *generalized transduction*, and has been found to occur in *E. coli, Pseudomunas, Proteus, Shigella* and many other types of bacteria. It is believed that, prior to cell lysis by phage, the host cell DNA breaks up into small fragments, and that it is one of these small fragments which comes to be included within the phage capsid. Sometimes both phage and bacterial DNA are packaged together, but often only a piece of bacterial DNA is packed in the capsid. Since these new phage particles are defective, having few if any phage genes, they do not lyse bacteria which they infect but rather endow them with new bacterial genes. Such transduction may prove to be *abortive* if the new genes are not integrated into the genome of the recipient cell. Nevertheless, the genes will function as a sort of temporary plasmid, bestowing on the recipient bacteria transient genetic properties.

In contrast to *general transduction* we will now discuss *specialized* or *restricted transduction*, which differs from the former in that particular bacterial genes are invariably transferred. This is believed to result from temperate phage such as lambda which insert into the bacterial chromosome at specific loci. There is a small but persistant chance that when lambda is excised from the bacterial genome the excision is defective, and a portion of the bacterial genome is cut out along with or instead of the phage genome. It follows that only those genes immediately adjacent to the prophage DNA are excised and transferred in specialized transduction.

Besides being an important mechanism for accomplishing variation in the bacterial genome, transduction has proved to be an enormously important technique in genetical research, since it enables single genes to be donated to host bacteria under rigorously controlled conditions. This approach, together with interrupted mating during conjugation, have pushed our knowledge of bacterial genetics far beyond our current understanding of the genetics of higher organisms.

3.1.3 Arrangement of Genes on the Bacterial Genome

We have already commented on the general non-random distribution of mapped loci on the genome of *E. coli* K12. There are one or two additional points to be made. The first is that some genes exist as multiple copies, and these copies are most frequently clustered together. *E. coli* has up to six copies each of the genes for 16S and 23S ribosomal RNA, and these are linked in map positions 83, 85 and 88 minutes. So the linkage is close, although many other loci are interspersed between. It does seem, however, that the 16 and 23S genes are each very closely linked in pairs, with probably only a short spacer between. There is also evidence that the 5S ribosomal RNA genes are multiple, with up to 10 copies, and some at least of these are close to the 23 and 16S loci.

Numerous genes coding for ribosomal proteins also exist, and many of these are clustered around map position 72 minutes. It is plausible to suggest that these ribosomal protein genes are clustered on grounds of transcriptional efficiency.

But a second type of gene clustering is also evident in the bacterial genome, and that is the close positioning of genes which do not code for the same or similar molecules but code for enzymes involved in the same metabolic pathway. The *lac* operon to be discussed later in this chapter is such an example, and there are many others. These groupings are therefore examples of functional clustering of genes, no doubt largely explained by the frequent mechanism of co-ordinate transcriptional control by operons, so marked a feature of pro-karyotic gene expression when it is compared with control systems in eukaryotes.

3.1.4   Control of Gene Expression in Bacteria

TRANSCRIPTIONAL CONTROL—POSITIVE AND NEGATIVE

When an RNA polymerase molecule traverses a gene and catalyses the pro-duction of an RNA molecule, the gene is said to be transcribed. Since even bacteria have some thousands of different genes, some only appropriate to very special metabolic situations, the cell does not transcribe all genes all of the time. Indeed this aspect of cellular economy is one of the most fascinating yet tantaliz-ing aspects of biology and even in bacteria we are far from a complete under-standing of how it is engineered. Transcriptional control mechanisms are styled positive or negative. *Positively* controlled mechanisms are those in which the gene is not transcribed without the assistance of a regulatory molecule. This molecule, when present, positively 'turns on' transcription of that gene. Negative control, on the other hand, is one in which the gene sequence can be transcribed in its native state. When it is 'off' it is being repressed by the presence of a specific regulatory molecule, and in normal metabolism that molecule must be removed to permit transcription of the gene. This, then, is the act of derepression and such genes are said to be *negatively* controlled.

As has been said earlier, a predominant feature of bacterial transcription is the co-ordinate control of genes whose products share common metabolic pathways. Although not all bacterial genes are known to belong to such groupings, the majority probably do, and we will limit our discussion to these fascinating assemblies of linked genes. Where the gene group is situated in one place on the genome and comes under the influence of a single regulatory mechanism, it is termed an *operon*, and such operons may themselves be regulated by positive or negative control.

The Lac Operon

By far the best known operon is that which controls enzymes involved in the metabolism of lactose, the *lac* operon, which is negatively controlled. We will therefore begin our discussion of operons by looking closely at this example. Not only is the '*modus operandi*' of the *lac* operon now very fully understood but its entire control region has been sequenced (see Fig. 3.3). Its original discovery stemmed from the observation that enzyme *induction*, the stimulation of production of a particular protein by the addition of inducer molecule to a culture, was co-ordinate in the case of three enzymes involved in lactose metabolism. In the absence of inducer none of the enzymes was synthesized;

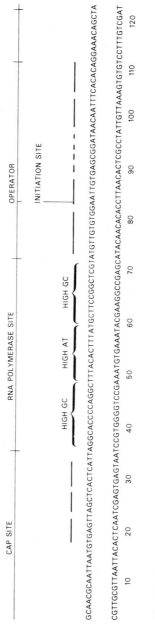

Fig. 3.3 The nucleotide sequence of the *lac* regulatory region. (From Dickson R. C. *et al*. (1975) *Science* **187**. 27.)

with the addition of inducer, all enzymes appeared. The phenomenon of co-ordinate enzyme induction was explained by Jacob and Monod in 1961 on the basis of their model of the operon. Addition of inducer was postulated to result in removal of a specific regulatory protein from the operator gene, resulting in derepression of the entire operon and facilitation of the reading of the structural genes by polymerases. The repressor was assumed to be the product of a specific regulatory gene and the act of induction to be accomplished by binding of inducer to repressor. Since the latter was assumed to be allosteric (i.e. to undergo an alteration in shape when associated with one molecule, this conformational change giving its alternative active site new affinities) association of inducer with repressor would lead to derepression of the operator gene and of the operon.

Figure 3.4 gives a diagrammatic representation of the *lac* operon and its mode of operation. Genes are arranged in the order I–P–O–Z–Y–A, the last three being the structural genes which code for enzymes of the lactose pathway. We will now proceed to list the definitions and characteristics of the types of genes and molecules involved in this fascinating genetic assembly.

(1) *Regulator gene I.* This gene, the first in the sequence, codes for the repressor protein. Mutants defective in this gene region either cannot effectively repress the operon and all the enzymes are synthesized at all times, (such cells are said to be constitutive mutants) or cannot derepress the operon and so can never synthesize the relevant proteins.

(2) *Promoter gene P* was first identified in 1966. This gene promotes transcription by providing a site at which RNA polymerase molecules can bind and so initiate transcription of the genes in the operon. The affinity of the polymerase molecule for P is also dependent on the presence of CAP protein on the CAP sequence which marks the first part of P.

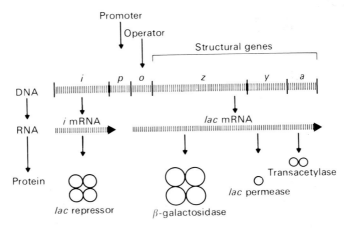

(a)

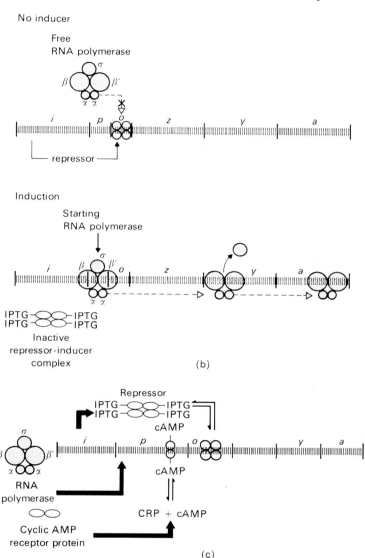

Fig. 3.4 (a) Diagram representing the *lac* operon; (b) Diagram representing regulation of the operon in the absence and presence of inducer; (c) Diagram representing the way in which cyclic AMP regulates the operon. (From De Robertis E. D. P. *et al.* (1975) *Cell Biology*. W. B. Saunders, Philadelphia, U.S.A.)

(3) *Operator gene O*. This is the gene to which the repressor binds. When inducer associates with repressor, the latter loses its affinity for O, and the polymerase is then free to traverse O and the structural genes beyond it. In other words, the operon is then derepressed.

(4) *Z gene* is the gene which determines the structure of the enzyme β-galactosidase, which hydrolyses lactose to galactose and glucose. This enzyme is a tetramer, and each subunit has a molecular weight of about 135 000.

(5) *Y gene* is the gene which determines the structure of a permease enzyme which facilitates the entry of lactose into the cell. The protein has a molecular weight of about 30 000 and specifically binds galactosides.

(6) *A gene* is the gene which determines the structure of a third protein, thiogalactoside transacetylase. The precise function of this protein is unknown and it does not seem to be essential for lactose metabolism. It is a dimer, each subunit having a molecular weight of 32 000.

(7) *Terminator sequence T* lies at the end of the A gene and ensures termination of transcription rather than read-through by the polymerase molecule.

(8) *Repressor protein*. This protein has been recovered from operator genes, and is a tetramer of four identical subunits and a combined molecular weight of 150 000. As mentioned earlier, it has separate binding sites for the operator gene and the inducer molecule, and, as an allosteric protein, loses its affinity for O when associated with inducer.

(9) *CAP protein*. This protein, originally styled as the catabolite gene activator protein, is known to be a dimer, each subunit being about 22 000 daltons in size. It also has a strong affinity for cyclic AMP and is often referred to as the cyclic AMP binding protein. An interesting region of two fold symmetry is to be found in the *lac* control sequence at the CAP attachment site, and it seems likely that CAP may recognize the site by its three dimensional structure. Only when CAP protein plus cyclic AMP bind to the CAP attachment site is it easy for the RNA polymerase to attach to the promoter sequence. In the absence of cyclic AMP, RNA polymerase binds to the promoter with only 5% of its normal efficiency, as determined by the level of induction of β-galactosidase enzyme in these conditions.

This then is a summary of how a negatively controlled operon works. We will now look at some other types of gene regulatory systems found in bacteria.

The arabinose operon—positive and negative control

This operon consists of a row of three structural genes, gene A coding for L-arabinose isomerase, gene B for L-ribulokinase, and gene D for L-ribulose 5-phosphate 4 epimerase. As shown in Fig. 3.5 these genes are arranged in the order D–A–B, with gene I, an initiator gene, and gene O, the operator gene, following. A sixth gene C, the regulator gene, is some distance away from the operator. Recent evidence indicates that the product of gene C is a protein which displays *alloteny*, that is that it can act as either repressor or activator, depending on whether it is bound to the inducer L-arabinose. When L-arabinose is scarce the regulatory protein acts as a repressor binding to O and preventing transcription, but when L-arabinose is abundant and conjugates with the C protein, the complex alters its binding allegiance and attaches to I. This greatly

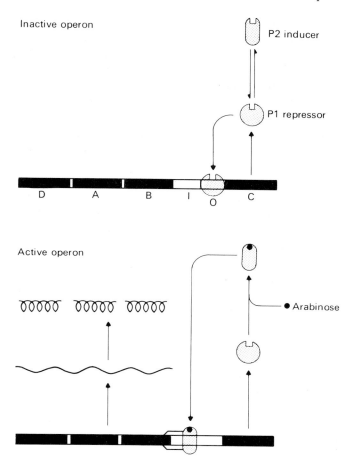

Fig. 3.5 Repressor-inducer model for control of the arabinose operon. The regulator gene araC codes for a repressor protein, P1, which binds at the operator (araO) to switch off the operon. The repressor is in equalibrium with an alternative protein conformation, P2, which acts as an inducer; but in the absence of arabinose P1 is the predominant state so the operon is inactive (upper). Addition of arabinose converts P1 repressor to P2 inducer; the inducer acts at the initiator site araI to switch on the operon. The molecular basis for its interaction is not defined but it is shown here as interacting with RNA polymerase to assist the initiation of transcription. Model of Englesberg, Squires and Meronk (1969). (From Lewin B. (1974) *Gene Expression Vol. 1.* John Wiley & Sons, Chichester.)

facilitates transcription of the operon in that, although some transcription occurs with the removal of C protein from O, it is not maximized until the C protein/inducer complex binds to I. This operon can therefore be said to function by both positive and negative control.

The arginine system—dispersed genes

The metabolic conversion of glutamic acid to arginine requires the assistance of eight enzymes, but the genes for these proteins map at eight different loci on the *E. coli* genome (see Fig. 3.6). But, in spite of their wide separation on the genome, the expression of all eight genes is controlled in parallel, each being subject to repression by the product of a single regulator gene (which is not linked to any of them) together with the co-repressor arginine. Interestingly, there are many more molecules of arginine repressor protein recoverable from a cell than lactose repressor, actually about 20 times more, and this no doubt reflects the fact that the same molecule must associate with eight operator genes instead of one, each structural gene requiring its own operator sequence. It is also pertinent to note that in this system, control of gene expression, although in parallel for the eight genes, is not strictly co-ordinate, since the levels of activity of the eight genes seem to be independently variable. Such a series of separate loci under common control of one regulator gene has been styled a *regulon* and it presents a fascinating anticipation of gene control in eukaryotes, where such effectors as hormones are known to modify the expression of many genes widely dispersed in the genome.

THE ROLE OF RNA POLYMERASE

Highly purified bacterial RNA polymerase contains the following polypeptides β^1, β, σ, α and ω. The enzyme complex is readily dissociated into the core enzyme (lacking σ) and the sigma factor (σ), the intact complex being termed holoenzyme. Sigma factor is not essential for all enzyme activity, since core enzyme will effectively transcribe calf thymus DNA. The ω chain also seems to be dispensable, and may not be an integral part of the complex. Core enzyme can therefore be expressed as $\alpha_2\beta\beta^1$. Sigma factor has the important role of restricting initiation of transcription to the promoter, and further, of ensuring affinity for particular promoters, as in the preferential synthesis of early genes in phage transcription.

In addition to sigma, a number of other proteins have been found to associate with bacterial polymerase and modify its affinities. Some of these accomplish selective transcription by ensuring increased affinity for certain promoters, others, as in the growth of phage lambda, alter the ability of the enzyme to recognize or answer to a termination sequence. This latter strategem can result in the enzyme reading through a terminator and thus transcribing further sequences downstream.

The ability of certain bacteria to sporulate also provides an interesting insight into the use of RNA polymerase. Sporulation amounts to a form of differentiation for the bacterial cell, although there is evidence that most vegetative genes are not turned off during sporulation but other new sequences are additionally expressed. In *Bacillus subtilis* a special sporulation-specific polypeptide, δ has been found. By combining with RNA polymerase core

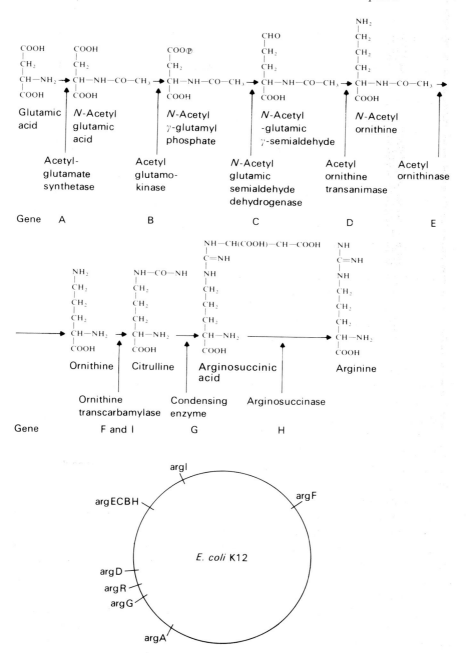

Fig. 3.6 The arginine biosynthetic system of *E. coli* K12. Strains B and W are similar but lack argF, so that OTC enzyme is produced only by argI. (From Lewin B. (1974) *Gene Expression Vol. 1*. John Wiley & Sons, Chichester.)

enzyme in varying amounts this protein seems to alter the promoter affinities of the polymerase. This then appears to be a specific control mechanism whereby selective gene expression is modulated in prokaryotes during sporulation.

3.1.5   DNA Replication in Bacteria

The bacterial genome is a circle of double stranded DNA, and its replication depends on the activity of the enzyme DNA polymerase. As in eukaryotes, synthesis is semi-conservative, and is now believed to be structurally discontinuous in both strands.

But, contrary to the eukaryotic model, DNA replication in bacteria is *continuous* in a growing culture, so there is no limitation to an S phase of the cell cycle. Moreover there is only one initiation point, from which replication occurs in both directions round the circle. There is also evidence that the initiation point for DNA replication is close to or part of the attachment site between the bacterial genome and the cell membrane. Growth of the membrane, or at least separation of two new initiation points on the membrane, seems to be part of the process of separating the new chains and ensures that each daughter cell receives at least one copy at cell division.

An additional point should be made about bacterial DNA replication, and that is that in a rapidly growing culture, a fresh round of DNA replication often begins at the initiation site before the previous round is complete. This leads to the slightly complex picture illustrated in Fig. 3.7, where three separate sets of replication forks can be distinguished. It is of interest to consider that genes located in portions of the genome which replicate early will normally be more abundant in the cell than genes which replicate late. The replication point of the *E. coli* genome lies close to 86 minutes (see Fig. 3.1) and termination close to 35. If the gene map is folded over between these points some symmetry of clustering and distribution is apparent and it is tempting to believe that gene location is in some way correlated with the amounts of gene products which are actually required in the economy of the cell.

3.2   VIRUSES

Viruses encompass a very varied group of particles, all being obligate intracellular parasites. Although they can survive outside cells they are entirely dependent on living cells for their replication. It is therefore clear that viruses cannot be a precellular form of life in evolutionary terms, but should rather be viewed as pieces of cellular genetic material which have gained some degree of individual autonomy. Viruses which infect plant, animal and bacterial cells differ considerably, and viruses are frequently classified according to their host cell type. They vary in their genetic and morphological complexity, but all

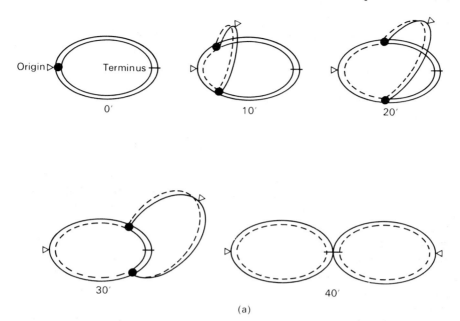

Fig. 3.7 a Diagrammatic representation of the genome of *E. coli*. DNA replication begins at a fixed site (the origin) on the 'chromosome' (actually a topologically circular double helix of DNA, here drawn as two parallel lines representing the two single strands of the helix). The black dots represent the positions of the replication complexes. Newly synthesized of DNA are shown as dotted lines. Replication proceeds at a constant rate until the two replication forks meet at the opposite site from the origin. This point, the terminus, is shown as a bar. The whole process takes about 40 min. (in most growth media at 37°C).

subscribe to a structure which comprises a genome of one type of nucleic acid, DNA or RNA, complexed to protein, the latter often forming a coat or capsid round the genetic material. The geometry of virus structure is often either a symmetrical cubic structure with 20 faces, i.e. an icosahedron, or a helical rod, although a few are more complex. Individual protein units in the capsid coat are referred to as capsomeres.

Not only do viral genomes vary in the type of nucleic acid, DNA or RNA, but also as to whether this nucleic acid is single or double stranded. It is intriguing to consider the varied approaches to intracellular replication adopted by different types of virus, and Fig. 3.8 illustrates the existence of six separate strategies. Thus the double stranded DNA viruses produce messenger RNA (termed +mRNA in the accepted convention) by the use of a normal RNA polymerase in the usual way (actually some viruses use their own special RNA polymerase, sometimes included in the capsid: but most depend on cellular RNA polymerases). The single stranded DNA viruses, as seen in the

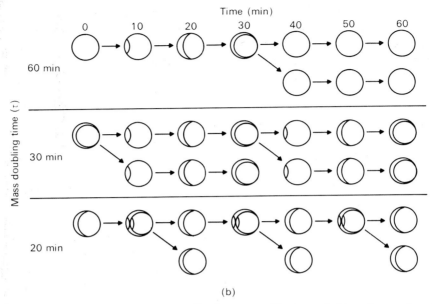

Fig. 3.7 b Pattern of chromosome replication over a 60 min. period in cells growing at different rates. The double helix is represented this time as a single line. In every case, each pair of replication forks take 40 min. to move around the chromosome from the origin to the terminus. A new round of replication is initiated at intervals equal to the mass doubling time of the cells, whether or not the previous round has been completed. (Both figures from Donachie W. D. *et al.* (1973) *Symp. Soc. Gen. Microbiol.* **23**, 9.)

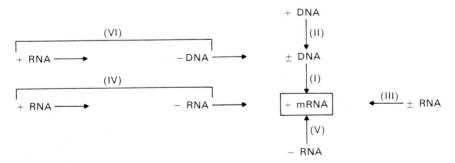

I Double-stranded DNA viruses	V Single-stranded RNA genome
II Single-stranded DNA viruses	complementary in sequence to mRNA
III Double-stranded RNA viruses	VI Single-stranded RNA genome with a
IV Single-stranded RNA viruses,	DNA intermediate in their growth.
mRNA identical in base sequence to	
virion RNA	

Fig. 3.8 Composite diagram of pathways of mRNA synthesis used by various classes of viruses, mRNA defined as +RNA. (From Pennington T. H. & Ritchie D. A. (1975) *Molecular Virology*. Chapman & Hall. London.)

figure, produce double stranded DNA by DNA replication prior to their transcription in the cell. Double stranded RNA viruses simply transcribe +mRNA from their own negative RNA strand, and no DNA intermediate is required in their replicative cycle. Similarly the two groups of single stranded RNA viruses, groups IV and V, require no DNA intermediate but transcribe messenger RNA from a negative RNA strand. Thus, all these three groups of viruses have accomplished a replicative cycle which completely omits DNA as a genetic material! The last group of viruses are perhaps even more fascinating, since they do have a DNA intermediate, but this DNA is synthesized from their own +RNA strand by an RNA dependent DNA polymerase, a process called reverse transcription. These type VI viruses then require replication of this DNA to give a double stranded DNA which then transcribes normal message in the usual way.

Another aspect of virus biology which we should briefly discuss before coming to consider individual viruses is the fate of the viral genome in the cell. In particular, the relationship of the viral genome to that of the host cell must be mentioned. Some viruses seem to be invariably virulent and, on entering the cell either as intact virus or viral genome, proceed to harness the cellular machinery for their own replication. Following viral replication, the cell dies and breaks open (lysis) and the newly produced virions are released. Some animal viruses have a variant of this cycle in which the host cell does not die and there is a steady release of virus from the cell. Such viruses are often released from the cell surface together with an enclosing piece of the cellular plasma membrane. Viruses causing this sort of infection are sometimes referred to as slow viruses. During a productive infection of a cell by virus, some copies of the viral genome may actually become inserted into the cellular genome, either by recombination with the cellular DNA or by some other process. Such cells may not betray the presence of the integrated virus in any way, but others may become altered in their metabolism or structural character, and are then said to be 'transformed'. Clearly the induction of cancer is one type of transformation that might, on occasions, be attributable to viral integration.

What is an occasional event during productive infection can itself be a complete alternative to such infection for certain virus, especially some phages. Such viruses, on penetrating the cell, become immediately integrated and no productive infection of virus ensues. Entire populations of eukaryotic cells or bacteria may therefore possess integrated viral genomes in their DNA without betraying its presence in any way. Such integrated virus is often termed provirus or prophage. In bacteria, where the phenomenon is not uncommon, phage which become integrated in this way are known as temperate phage (in contrast to virulent phage) and the infected bacteria are said to be lysogenic. The stable relationship between a temperate phage and its lysogenic bacterial host may break down, especially under the influence of ultraviolet radiation, resulting in the alternative of viral production and cell lysis. These alternative states of

Table 3.2a Summary of viron properties of some major DNA virus groups.

Virus group	Genome	Nucleocapsid symmetry	Transcriptase	Number of different polypeptides
Filamentous bacteriophages (fd)	S,[a] C,[b] 2[c]	Helical	−	2
Spherical bacteriophages (ϕX174)	S, C, 1.7	Cubic, 12 capsomers	−	5
Bacteriophages with non-contractile tails (λ)	D, L, 32	Complex	−	12
Bacteriophages with contractile tails (T2)	D, L, 130	Complex	−	30
Papovaviruses	D, C, 3–5	Cubic, 72 capsomers	−	6
Adenoviruses	D, L, 20–25	Cubic, 252 capsomers	−	13
Herpesviruses	D, L, 100	Cubic, 162 capsomers	−	30
Poxviruses	D, L, 160	Complex	+	> 30

[a] S, single-stranded; D, double-stranded.
[b] C, circular; L, linear.
[c] molecular weight $\times\ 10^{-6}$ daltons.

integration and non-integration apply also to some plasmids such as sex factor, as we discussed earlier in relation to the genetics of bacteria and plasmids.

In this chapter we will consider only a few chosen types of virus in detail, but Table 3.2a and b gives a summary of the properties of the main groups of viruses. For those interested in viral implication in tumour formation, an interesting but highly complex topic, the point should be made that whilst one group of RNA-containing viruses are referred to as tumour viruses, many different groups of DNA viruses include some examples which are oncogenic, that is, they are involved in tumour formation.

The examples of viruses which we will now consider are chosen partly because they are important in terms of their genomic material, partly because a great deal of information has been accumulated about them experimentally. Certainly some viruses are known and understood genetically almost to completion, a situation which compares startlingly with our ignorance of eukaryotic genetics.

3.2.1   Simian Virus 40

This virus, termed SV40 for short, is a now famous example of the papovaviruses, small animal viruses with double stranded DNA as their genome. SV40 was isolated originally from cultures of cells from the African green

Table 3.2b Summary of viron properties of some major RNA virus groups. (Both Table 3.2a and b from Pennington T. H. & Ritchie D. A. (1975) *Molecular Virology.* Chapman & Hall, London.)

Virus group	Genome	Nucleocapsid symmetry	Envelope	Trans-criptase	Number of different polypeptides
Spherical bacteriophages (Qβ)	S,[a] $+$,[b] 1[c]	Cubic	—	—	2
Spherical plant viruses	S, $+$, 1	Cubic	—	—	1
Filamentous plant viruses (TMV)	S, $+$, 2	Helical	—	—	1
Picornaviruses (poliovirus)	S, $+$, 2.6–2.8	Cubic	—	—	4
Togaviruses	S, $+$, 4	Unknown	$+$	$-$	3
RNA tumour viruses	S, $+$, (10)	Unknown	$+$	$+$ (reverse)	7
Rhabdoviruses	S, $-$, 4	Helical	$+$	$+$	5
Paramyxoviruses	S, $-$, 7	Helical	$+$	$+$	6
Myxoviruses	S, $-$(3–5)	Helical	$+$	$+$	6
Reoviruses	D, (15)	Cubic	—	$+$	7

[a] S, single-stranded; D, double-stranded.

[b] Polarity of genome strand; $+$ strands are mRNA-like.

[c] Molecular weight $\times 10^{-6}$ daltons; viruses with fragmented genomes have this figure in parenthesis.

monkey, and the virus seems to occur commonly in wild monkeys. As seen in Fig. 3.9 the electron microscope reveals it to be a typical small icosahedral particle, about 40 nm in diameter. Its genome consists of a closed loop of double stranded DNA of mol wt about 3.4×10^6.

INFECTIVE CYCLE

Alternative patterns of infection are an important feature of this virus. In cell cultures of African green monkey kidney the virus replicates productively and cells are lysed with release of the new virions. Cell lines which undergo this

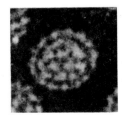

Fig. 3.9 Electron micrographs of SV40 virus. Photo kindly supplied by Professors T. J. Kelly and D. Nathans.

cycle of virus production are said to be 'permissive'. In certain other cell types, however, an alternative course of events is found. If SV40 is grown in mouse cells, for example, only some of the viral proteins are made and no complete virus is released, but many of the affected cells come to display new heritable growth properties, that is, they are transformed. There are good grounds for believing that transformed cells have one or more copies of SV40 genome integrated in their own DNA, and that often copies of the viral DNA are inserted in more than one site in the chromosome set of the host cell. It is also interesting to note that cellular transformation by SV40 DNA has been accomplished with pieces of DNA which are less than the complete virus genome.

THE SV40 CHROMOSOME
This virus is of the greatest relevance to the main topic of this book chiefly because it is an example of a virus with a chromosome composed of chromatin. Such structures are sometimes termed minichromosomes. As seen in Fig. 3.10 the electron microscope reveals that this chromosome consists of a string of beads and, on closer examination, such beads have proved to be nucleosomes composed of the conventional 8 molecules of histone conjugated with the DNA. Not only is the SV40 genome a true chromosome in the cell, but this chromatin complex persists even when the genome is packaged inside the capsid coat. Since SV40 virus is predominantly found in host cell nuclei and its productive cycle seems to be completed there, the association of histone with its DNA is perhaps not too surprising. But another interesting observation stems from the presence of histone on SV40 DNA. That is that this histone is a host cell-coded molecule, yet is included in the structure of the completed virion. It is also thought likely, though not as yet proved beyond doubt, that some SV40 coat proteins are cellular proteins, implying that the virus uses some cellular components to achieve its own architectural form.

SV40 DNA REPLICATION
The replication of the closed loop of SV40 DNA occurs by a process analagous to that operative in bacteria. One initiation point is present and replication proceeds in both directions from that point at an approximately equal rate, terminating at the opposite side of the circle after each point has traversed through 180 degrees (see Fig. 3.11). By eukaryotic standards SV40 DNA replication is very slow, requiring between 5 and 25 min. to complete. An extrapolation from eukaryotic DNA replication would suggest a time of only 25 s. It is thought that the discrepancy in time may be related to the fact that the SV40 genome is enzymatically nicked and sealed many times during replication since the DNA is in the form of a superhelix which has to be unwound to permit replication.

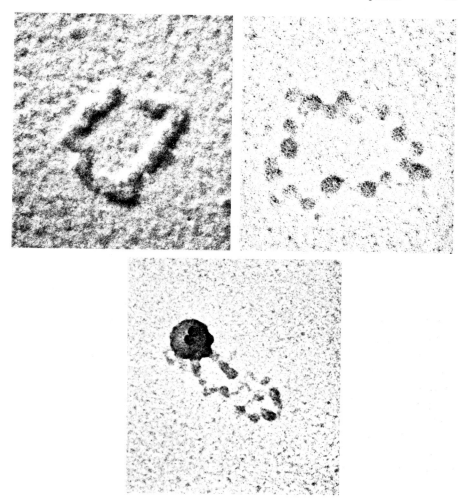

Fig. 3.10 SV40 mini-chromosomes: Electron micrographs of SV40 DNA–histone complexes isolated from infected cells and fixed in 0.15 M NaCl (top left) or 0.01 M NaCl (top right). Similar complexes can be isolated from SV40 virions. At the bottom is a complex 'partially extruded' from a virion following brief treatment at pH 9.8. (Photographs kindly supplied by Professor T. J. Kelly and D. Nathans. (From *Adv. Virus Res.* (1977) **21**, 86.))

SV40 TRANSCRIPTION

A point made in the introduction to this chapter is well illustrated by SV40 transcription, namely the highly economical use of its genome. This economy is borne out by two separate aspects of its transcription, viz: both strands of the DNA, the L or minus strand and the R or plus strand, are transcribed into mRNA; one particular message overlaps another so that the same part of the

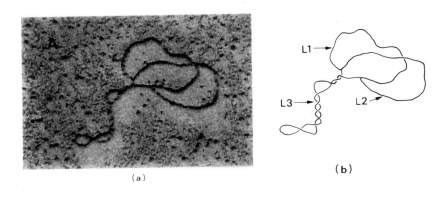

(a)

(b)

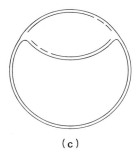

(c)

Fig. 3.11 SV40 DNA replicating intermediate: (a) electron micrograph of SV40 replicating intermediate; (b) Interpretive drawing—replicative intermediates contain two forks, three branches, and no visible ends. Two of the branches (L1 and L2) are of equal length and represent the replicated segments of the molecule. The third branch (L3) contains superhelical twists and represents the unreplicated portion of the molecule. (c) Summary of the basic structural features of SV40 replicating intermediate. Both parental strands are covalently closed. The daughter strands are held to the parental strands by hydrogen bonds alone. Both daughter strands grow by a discontinuous mechanism, and there may be single-stranded regions (in the trans configuration) at each fork. (From Kelly T. J. & Nathans D. (1977) *Adv. Virus. Res.* **21**, 86.)

same strand is used to produce two separate classes of mRNA. Figure 3.12 illustrates this pattern of transcription, and also indicates that one type of transcript is termed 'early', another 'late'. This implies that the early mRNA is transcribed first, early in the infective cycle, and the other more stable mRNA is transcribed 'late' in the cycle, and indeed not until DNA replication itself has begun.

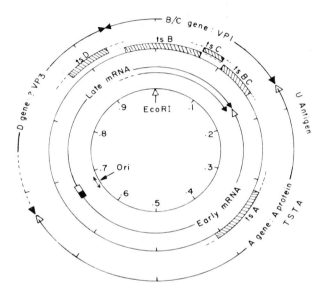

Fig. 3.12 Summary map of the SV40 genome. (From Kelly T. J. & Nathans D. (1977) *Adv. Virus Res.* **21**, 86.)

Another interesting aspect of SV40 transcription is that, like many eukaryotic messenger RNAs, the major SV40 mRNA, which is known to code for a viral protein, is comprised of at least two separate pieces transcribed from separate sections of the genome. A leader sequence of up to 200 bases in length is attached to the 5' end of the main 16S mRNA but on the genome the two relevant coding sequences are a long distance apart. In fact very recent evidence indicates that in SV40, and also in another animal virus, adenovirus, such leader sequences may be shared by a number of different viral messages, and that the leader sequences themselves are mosaics of pieces of RNA whose coding sequences are not contiguous. Figure 3.13 illustrates the surprising findings in adenovirus. Since the entire DNA sequence of SV40 has now been determined, a feat, one should say, of astounding technological application (ϕX174 has also been entirely sequenced) our knowledge of SV40 coding strategy is very considerable.

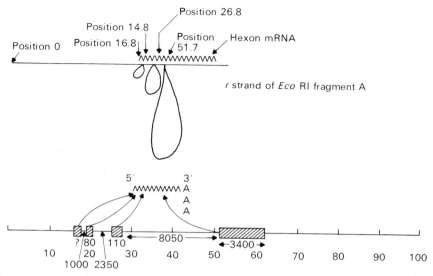

Fig. 3.13 Formation of adenovirus 2 in RNA for the Hexon gene showing interrupted leader sequences and main gene are separated by introns (see discussion also in Chapter 5). (From Sambrook J. (1977) *Nature* **268**, 101–4.)

3.2.2   T4 phage

All the other viruses which we will discuss here are phages. This is not because other animal and plant viruses are not interesting in terms of their DNA, but simply because phages have been very intensively investigated and come in many different forms. Knowledge of phage genetics is therefore at an advanced stage.

    Of all the phages that infect *E. coli* the T-even phages T2, T4 and T6 are the largest and most dramatic. None of these is a temperate phage, since the bacterial chromosome is itself broken down early in infection and so viral production and lysis are inevitable. Much early work was carried out on T2, but in recent years work on T4 has intensified and its genetics are now fairly fully understood.

T4 ORGANIZATION

The T-even phages are very big viruses with a complicated and curious structure. The head is of the usual icosahedral structure but to it is attached a large tail made up of a thick and hollow mid-piece, a base plate, and a set of lengthy tail fibres (see Fig. 3.14). It is clear that quite a few different proteins are going to be involved in putting a structure of this sort together, and T4 DNA probably codes for up to 150 different proteins, and the capsid itself includes some 30 different polypeptides in its structure. The genome of this phage is a long

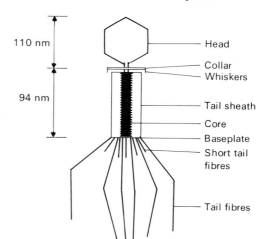

110 nm

94 nm

Head

Collar
Whiskers

Tail sheath

Core

Baseplate

Short tail
fibres

Tail fibres

Fig. 3.14 Diagram of the virion of phage T4. (From Pennington T. H. & Ritchie D. A. (1975) *Molecular Virology*, Chapman & Hall.)

molecule of double stranded DNA comprising 166 000 base pairs. The molecule is not a closed circle although, for reasons which we shall discuss shortly, a circular gene map can be drawn.

INFECTIVE CYCLE

As we have mentioned above, the infective cycle of T4 follows only one possible programme which involves productive infection, cell lysis, and release of new phage. Only the DNA genome of the virus enters the host cell on infection, and the manner of entry of the DNA remains unclear, although contraction of the phage tail mid-piece is known to occur during the process.

T4 GENOME

A diagram of the T4 genome is shown in Fig. 3.15. Its circularity is explained by the fact that different individual phage particles have the circle broken at different places, so that different molecules begin and end at different points in the sequence. In any population the T4 genomes are therefore said to be *circularly permuted* with respect to one another. In addition, each molecule is *terminally redundant* so that the sequence at one end is repeated at the other, and each length contains a little more than one genome-equivalent of DNA (actually about 2% more).

The diagram of the T4 genome indicates a further remarkable feature of this molecule in connection with the actual order of genes, and that is that genes coding for functionally related molecules are clustered on the genome, so that there is a group of tail baseplate genes, a group of DNA replication genes, and so on. Some but probably not all of these clusters are transcribed as polycistronic messages, i.e. RNA molecules with a number of genes coded in tandem, from single promoters.

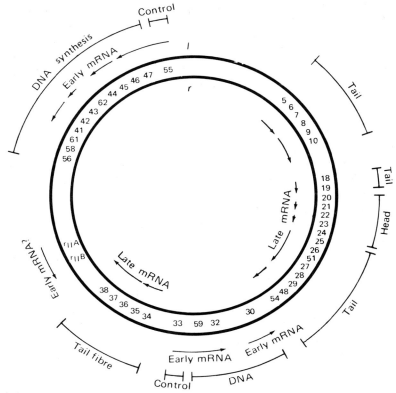

Fig. 3.15 Partial genetic map of phage T4 showing functional grouping of genes and pattern of transcription of some map segments (arrows). Numbers represent individual genes. (From Pennington T. H. & Ritchie D. A. (1975) *Molecular Virology*. Chapman & Hall, London.)

Another point to note about the DNA of T4 and indeed of all T-even phages is that it is characterized by the presence of an unusual base, 5-hydroxymethyl-cytosine, in place of the normal cytosine. This fact has greatly facilitated studies of DNA synthesis in T-even phages.

T4 TRANSCRIPTION
Although T4 does not transcribe much of its DNA in an overlapping manner, as SV40 does when making different messenger RNAs, it does utilize both strands of DNA as sense coding strands. Moreover, the feature of having distinct early and late transcription is even more evident in T4 than in SV40. The timing of T4 transcription has been divided into early, quasilate (a nasty word, in my view) and late, and each set of genes in these different temporal categories has distinct promoter regions. As we might expect, the early T4 genes code for many proteins which are themselves necessary for the transcription or transla-

tion of late ones. Although in fact all T4 genes are transcribed by the host RNA polymerase, there is evidence that late promoters on T4 can only be recognized by the host polymerase with the assistance of polypeptides coded by early genes.

T4 infection also affects cellular translation, since the phage codes for several species of transfer RNA. There may therefore be some translational control of phage protein production, perhaps at the expense of host protein synthesis.

3.2.3   Phage Lambda—Organization

This phage, like T4, consists of an icosahedral head with attached tail, but the latter is simpler than in T4 with no contractile mid-piece and only a single tail fibre (see Fig. 3.16). Lambda is the best known of the temperate phages, viruses which are episomes in that they may be integrated into the host cell genome, behaving just like a cluster of bacterial genes, or alternatively may indulge in autonomous replication with consequent cell lysis. And we should remember that infection of an *E. coli* cell with lambda, and its consequent integration and adoption of lysogeny, renders that cell immune to further attack by phage of the same type.

Lambda is a small phage with a genome of double stranded DNA, enough for at least 35 genes and of a molecular weight 3.1×10^7 daltons, i.e. some 46 500 base pairs. The genes and gene products of this phage have been well characterized and it is interesting to note that of the 35 genes in the genome, 20 are concerned with producing head and tail proteins, 9 with DNA replication and lysis, and 6 are regulatory. The question of whether the lambda genome is linear or circular is an interesting one, since molecules with both configuration can be isolated. It seems that infective lambda DNA is linear, but has *free cohesive ends* consisting each of a short length, 12 bases in fact, of single stranded DNA, each free end being exactly complementary to the other free end. A DNA ligase enzyme is able to convert linear lambda DNA into complete covalently closed circles, as shown in Fig. 3.17.

LYSIS OR LYSOGENY—HOW THE CHOICE IS MADE

The question of whether an individual viral genome, on entering a bacterial cell, will proceed to a lytic or lysogenic cycle seems to be a rather random one, and in a laboratory culture situation, some do one thing and some another. It is known to involve competition between two regulatory proteins, one termed repressor and the other CRO. Both of these proteins bind to the same control sites and influence the same promoters. Repressor protein prevents activity of genes involved in the lytic cycle, while CRO antagonizes lysogeny and permits lysis to proceed. The balance struck between the repressor and CRO protein seems to be straight competition for DNA binding sites, although it is somewhat complicated by the fact that each tends to repress its own synthesis to some extent.

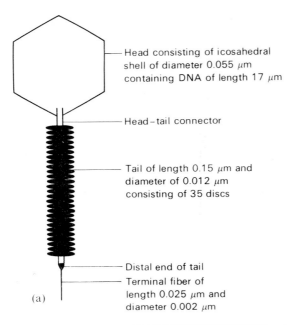

Head consisting of icosahedral
shell of diameter 0.055 μm
containing DNA of length 17 μm

Head–tail connector

Tail of length 0.15 μm and
diameter of 0.012 μm
consisting of 35 discs

Distal end of tail
Terminal fiber of
length 0.025 μm and
diameter 0.002 μm

(a)

(b)

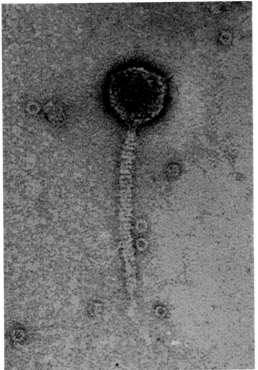

Fig. 3.16 (a)   Structure of
phage lambda virion. (b)
Electron micrograph of phage
lambda × 250 000. (From
Lewin B. (1977) *Gene Expression
Vol. 3*. John Wiley & Sons,
Chichester; (b) kindly supplied
by Dr A. Howatson.)

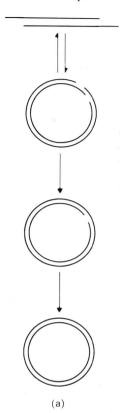

Fig. 3.17 (a)   Linear and circular forms of DNA. The linear form has two protruding single strand ends which can anneal with each other to form a circle with a break on each strand. A covalently closed circle can be formed when both breaks are sealed. (b) Cohesive ends of DNA. The upper molecule shows the sequence of nucleotides incorporated into the cohesive ends of DNA polymerase 1. The lower molecule shows the corresponding sequences of the protruding cohesive ends. Data of Wu and Taylor (1971). (From Lewin (1977) *Gene Expression Vol. 3*. John Wiley & Sons, Chichester.)

(a)

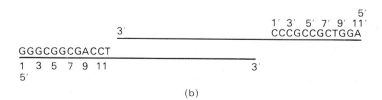

(b)

INSERTION OF THE PHAGE GENOME IN LYSOGENY

For many years there was considerable speculation about the relationship between the phage and bacterial DNA in lysogeny. It is now quite clear that the phage genome comes to be colinearly inserted into the bacterial chromosome and so is normally replicated with it by the DNA polymerase. There is a specific sequence in the *E. coli* chromosome which the lambda phage recognizes and no host functions appear to be involved in this sequence. (We should note that some phages other than lambda appear to be able to integrate at many sites, perhaps anywhere.) There is also a sequence in the lambda DNA which recognizes the bacterial site. The actual DNA linkage between phage and bacterial DNA is achieved by 'site specific recombination', probably as illustrated in Fig. 3.18. This process is known as the Campbell model of additive recombination. Lambda must be in circular form to permit site recognition and recombination, but following insertion the prophage DNA is of course linear.

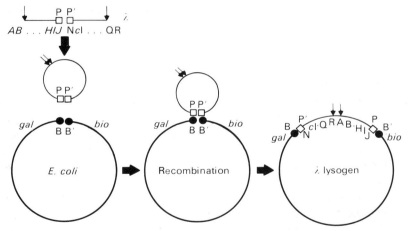

Fig. 3.18 The Campbell model for insertion of phage lambda into the *E. coli* chromosome. (After Smith-Keary P. F. (1975) *Genetic Structure and Function.* Macmillan, London.)

At a certain low frequency the lysogenic state is converted to the lytic state, and exposure of lysogenic cells to either mitomycin C (a drug which inhibits DNA synthesis) or ultraviolet light greatly enhances this change. This process is known as induction and involves excision of the prophage genome from its insertion site on the bacterial chromosome.

GENE TRANSCRIPTION IN PHAGE LAMBDA

As with phage T4, lambda genes are highly ordered and the regulated events of the lytic cycle are similar to those of T4, except that there is much less dramatic interference with host cell protein and DNA synthesis in the lambda lytic cycle than in T-even phage cycles. Figure 3.19 illustrates the clustering of lambda genes and the pieces of DNA which are known to code for them.

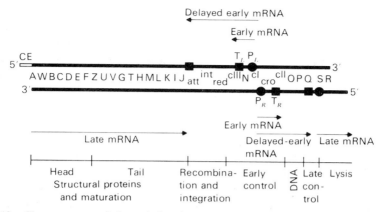

Fig. 3.19 Chromosomes of phage λ showing arrangement and functions of genes and pattern of transcription (arrows). Circles and squares indicate promotors and terminators respectively. CE denotes cohesive end. (From Pennington T. H. & Ritchie D. A. (1975) *Molecular Virology*. Chapman & Hall, London.)

3.2.4 Single Stranded DNA Phages

These phages are actually of two fairly distinct types, filamentous and icosahedral. Both have genomes of about the same size, and some eight or nine genes have been identified. Table 3.3 gives a summary of their characteristics. φX174, illustrated in Fig. 3.20 is the best known of the cubic shaped phages, and the F class of phages are single stranded filamentous viruses which adsorb to the tip of the bacterial F pilus.

φX174 infection involves conversion of the single stranded phage genome into a double stranded molecule soon after entry, followed by a normal lytic cycle. Viral mRNA synthesis is not divided into early and late species but the pattern of transcription is none the less interesting in that the different types of transcripts overlap one another and have a single common promoter site. The varied messages produced are the results of differing termination sites, some of which are not completely effective (see Fig. 3.21). Of the mRNA produced, the long molecules appear to be polycistronic so the products of the first genes transcribed are found to be much more abundant than those of the last.

Yet another astonishing transcriptional strategy is demonstrated by φX, this time the use of the same DNA sequence to produce two quite distinct proteins by a reading frame shift. A sequence of φXDNA has been found to represent a gene termed D, but within the latter part of D is another gene, E, and this gene is read in a different triplet reading frame from D. And, not to be outdone, another single stranded DNA phage, termed G4, has very recently been found to code three proteins from the same DNA by utilizing three different reading frames. Surely the ultimate in coding economy! Many of these features of code utilization are now coming to light with progress in sequencing, and

Fig. 3.20 Particles of ϕX174 phage, visualized in the electron microscope by the aid of platinum shadowing. Photograph kindly supplied by Dr I. Tessman, Pardue, U.S.A.

Table 3.3 Particles of single strand DNA phages

	Icosahedral (ϕX174)	Filamentous (fd)
Particle mass	6.2×10^6 daltons	14.6×10^6 daltons
DNA content	1.7×10^6 daltons (5500 bases)	1.9×10^6 daltons (6000 bases)
Protein content	4.5×10^6 daltons	12.7×10^6 daltons
Protein constituion	~ 60 copies of pF (48 000 daltons)	~ 2400 copies of pVIII (5240 daltons)
	~ 60 copies of pG (19 000 daltons)	< 4 copies of pIII (56 000 daltons)
	< 12 copies of pH (36 000 daltons)	
	? copies of pJ (9000 daltons)	
Particle dimensions	Icosahedron of diameter 250 Å	filament of 8500 $\times$ 60Å

Estimates for the mass of the phage particle and the content of DNA and protein are based on physical analysis of mature phages. The protein constitution is based upon the identification of viron components and the number of copies of each protein is calculated from the proportion of the protein mass it occupies; this is reasonably accurate for the fd phage but may be in error for ϕX174. (From Lewin B. (1977) *Gene Expression Vol. 3.* John Wiley & Sons, Chichester.)

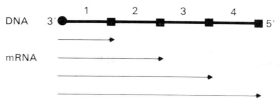

Fig. 3.21   Scheme for overlapping transcription of ϕX174 DNA. The four DNA segments are transcribed from a single promotor (○) with weak terminators (□) at the end of each segment. mRNA for segment 1 is consequently more abundant than mRNA for segment 4. (From Pennington T. H. & Ritchie D. A. (1975) *Molecular Virology.* Chapman & Hall, London.)

indeed the entire sequence of the 5375 nucleotides of ϕX174 has now been determined (see paper by Sanger G. M. *et al.* (1977) *Nature* **265**, 687–95).

Infection of a bacterium with filamentous single stranded DNA phage is distinct in a number of ways. Firstly, many of these phages have particular affinities for bacterial pili, the F phages for the F pilus. Entry of the phage involves the whole filament, DNA and protein coat, and also involves retraction of the F pilus to which the phage is attached. The precise manner of entry of the phage remains unclear. Another distinct feature is that, following infection, no cell lysis occurs but new phage is continuously released from the cell by extrusion through the cell wall. Infected cells even continue to divide and grow. So these phages are, in their lifestyle, analagous to the slow productive infections of some animal viruses.

In both ϕX and F phages, the parental replicating DNA must eventually produce single plus strand DNA for the genomes of new phage. This is thought to be produced by a rolling circle type of synthesis, the new strand peeling away from the double stranded parent molecule. The remarkable photograph shown in Fig. 3.22 is believed to illustrate this process. Figure 3.23 gives a general scheme for DNA replication in ϕXl74. Our discussion of ϕXDNA should perhaps conclude with the observation that the circle of DNA which comprises the genome of this virus is given tertiary structure by the presence of superhelical turns in its DNA, and a special enzyme, termed DNA gyrase, is responsible for accomplishing this spatial arrangement.

3.2.5   RNA Phages

As Figure 3.8 indicates, there are four distinct types of RNA viruses, one double stranded and three single stranded. We will only discuss one type here, namely RNA phages in which the genome is a single stranded RNA and in which the mRNA is identical in base sequence to virion RNA. Various phages have been identified as belonging to this group, of which phage Qβ is perhaps the best known. They are all remarkably small viruses, with linear genomes of about 4000 nucleotides coding for only 3 or 4 different proteins—a major coat protein, a minor coat protein, a gene coding for a replicase, which is one polypeptide

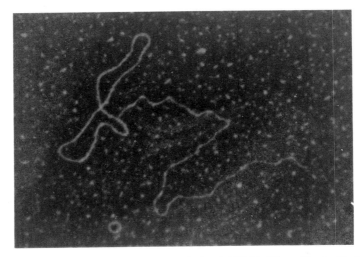

Fig. 3.22   Electron micrograph of a rolling circle of ϕX174. (From Koths K. & Dressler D. (1978) *Proc. Natl Acad. Sci.* **75**, 605.)

of a multi-unit replication enzyme for viral genome replication, and, in the case of Qβ but not f2, one other protein. Although single stranded, the genome is hydrogen bonded in many places to provide hairpin—like loops—see Fig. 3.24. Since the genome is itself the mRNA, it is not surprising to find that certain sequences in the genome are not translated, but serve as ribosome attachment sites. The specific folding of these regions determines ribosomal affinity and gives the phage a measure of translational control over which of its proteins are made in greatest quantity at specific times. Viral replicase protein, for example, is synthesized only early in infection and switched off later. Coat protein and A are made continuously but the former is always made in great excess. Indeed it is coat protein itself which is able to exert translational control over the formation of replicase molecules. When coat protein accumulates in the cell, six molecules of the coat protein associate with the initiation site of the replicase gene and prevent further ribosomes being instrumental in its translation. This is represented in Fig. 3.25.

As with filamentous DNA phages, these phages adsorb to bacterial sex pili, and have indeed been utilized as a sex pilus label (see Fig. 3.26) but, unlike the filamentous DNA phage, the genome of these RNA phages seem to enter the cell without the coat protein but with protein A.

3.2.6   Viroids

Viroids are pathogens of higher plants and exist only as self replicating uncoated RNA molecules. They are single stranded and circular, but the circle is almost certainly considerably base paired to yield a rod-like molecule with closed ends. Cutting of the circular structure has been found to result in loss of infectivity.

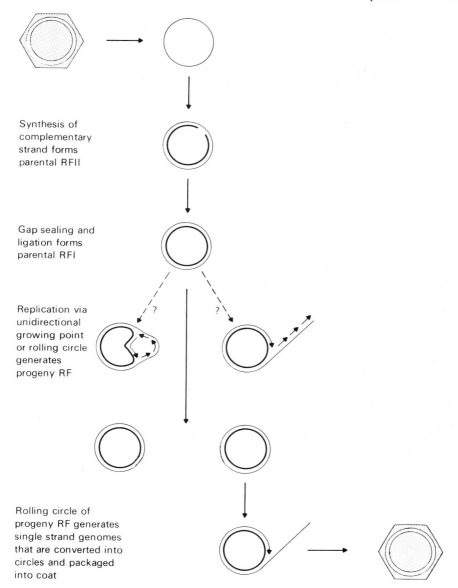

Synthesis of
complementary
strand forms
parental RFII

Gap sealing and
ligation forms
parental RFI

Replication via
unidirectional
growing point
or rolling circle
generates
progeny RF

Rolling circle of
progeny RF generates
single strand genomes
that are converted into
circles and packaged
into coat

Fig. 3.23   Model for replication cycle of φX174. (From Lewin B. (1977) *Gene Expression Vol. 3*. John Wiley & Sons, Chichester.)

Surprisingly, viroids do not seem to function as messenger RNA molecules nor interfere with the translation of other known messengers. It is even difficult to understand, on structural grounds, how they can associate with ribosomes, so they may entirely lack the facility for translation. Although it is too early to be

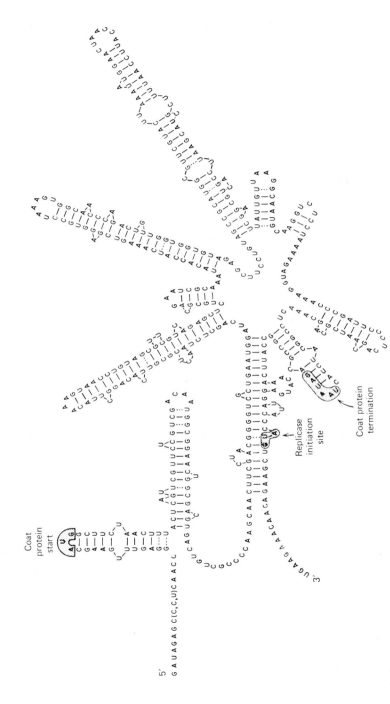

Fig. 3.24 The flower model for the secondary structure of the coat protein gene of MS2 RNA. The coat protein initiation codon lies at the end of a hairpin structure. The replicase initiation site is base to a conplementary sequence early in the coat protein gene. (Data from Min Jou W. *et al.* (1972) *Nature* **237**, 82. From Lewin B. (1977) *Gene Expression Vol. 3*. John Wiley & Sons, Chichester.)

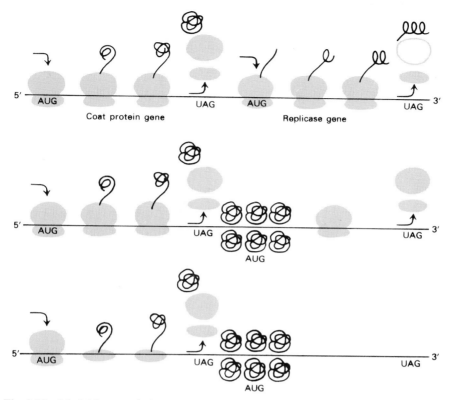

Fig. 3.25 Model for translation repression by f2 coat protein. Both the coat protein gene and replicase gene are under translation. When the concentration of coat protein becomes great enough, up to six molecules of coat protein bind to a region including the initiation site of the replicase gene: this prevents further ribosomes from attaching and ribosomes currently on the gene continue until they reach the termination codon. This clears the replicase gene of ribosomes but the coat protein gene remains under translation. (From Lewin B. (1977) *Gene Expression Vol. 3*. John Wiley & Sons, Chichester.)

at all certain, viroids may be a purely RNA parasite, but this leaves unexplained their ability to produce pathological symptoms in the host plant. One viroid has been entirely sequenced, namely potato spindle tuber viroid (PST V) and has been found to possess 359 nucleotides giving a molecular weight of 115 000. A suggested secondary structure for PST V is illustrated in Fig. 3.27.

3.3 PLASMIDS

Plasmids are genetic factors which enjoy an independent existence in cellular cytoplasm and are not themselves part of the normal genome of that cell. A cell

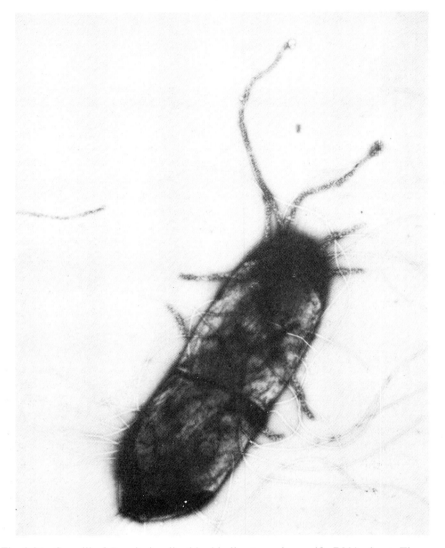

Fig. 3.26 Sex pili of *E. coli* visualized by binding to male-specific RNA phage. The phage particles are about 250 Å in diameter and bind along the length of the pilus. Photograph kindly provided by A. Lawn, data from Datta N., Lawn A. M. & Meynell E. (1966) *J. Gen. Microbiol.* **45**. 365.)

can normally survive without the presence of plasmids, although some plasmids may incidentally favour the survival of the host cell in some way. To date, all known plasmids have double stranded DNA as their genetic material and they are able to replicate in the cell independently of the replicative cycle of the host. As a matter of original definition a plasmid was not capable of being integrated

Fig. 3.27 Suggested structure for the potato spindle tuber viroid. Figure kindly supplied by Dr. H. J. Gross. (From Gross H. J. *et al.* (1978) *Nature* **273**, 203–8.)

into the cellular chromosome. If a factor was capable of such integration, at least in certain circumstances, it was styled an episome. Recently the two terms have been used more loosely and now the word plasmid, in general use, should not be taken to imply that integration into the host genome is not possible.

Although plasmids are not confined to bacterial cells, our knowledge of eukaryotic plasmids is flimsy and, with the exception of the interesting case of crown gall disease in plants, we will confine our discussion to plasmids of prokaryotic cells. It is of course amusing to speculate on the possible role of eukaryotic plasmids in disease and even as an explanation for the transposable genetic elements discovered in many eukaryotic cells (see p. 265).

Bacterial plasmids can be subdivided into three main groups, sex factor, drug resistance factors and colicin factors, and these we will now consider in turn.

3.3.1 Sex Factor and Bacterial Conjugation

In the late 1940's and early 50's, Joshua Lederberg and others discovered that genes could be transferred from one bacterium to another in a process which demanded physical contact between the differing bacterial strains. The process is termed conjugation and its discovery was quickly followed by the appreciation that such genetic transfer was invariably one-way, from a donor cell to a recipient cell. It was evident that different strains of *E. coli* existed, some strains having the ability to conjugate with other strains, and that the donor strains possessed a genetic factor which they passed on at conjugation. This factor was termed sex factor, and bacteria possessing it were said to be F^+ and those lacking it to be F^-. Actually, in the original experiments, it was the transfer of other bacterial genes along with sex factor that was detected, but it is now clear that such additional transfer does not invariably accompany sex factor transfer.

Sex factor has turned out to be a length of double stranded DNA of about 4.5×10^7 dalton, being about 2% of the length of the entire *E. coli* genome, with space for anything up to 100 genes (although only about a dozen are known). In the F^+ bacteria this factor is free in the cytoplasm and is transferred to F^- bacteria by conjugation of an F^+ with an F^-. Amazingly, it now seems that the sexual process in bacteria is entirely dependent on the possession of this plasmid and that strains or species of bacteria lacking it cannot conjugate (conjugation has been recorded in various species of *Escherichia*, and also in *Salmonella*, *Shigella*, *Serratia* and *Pseudomonas*. Indeed some of these species will conjugate together and exchange sex factor). This view of conjugation depending on possession of sex factor is underlined by the fact that the physical contact between conjugating bacteria involves a special structure, the F pilus, the proteins of which are coded for by the sex factor genome. Many bacteria possess pili in the normal course of events, but F pili, possessed solely by F^+ strains, are larger than and quite distinct from, normal pili. The precise role of the F pilus in conjugation remains unclear, although some speculation exists that the F factor DNA passes through it (thus its name of conjugation tube). This idea remains purely speculative.

Sex factor in F^+ bacteria is therefore an infective property. The F factor resembles a temperate phage in many of its characteristics and its stable relationship with the host cell, but of course it lacks the ability to provoke lysis in the host. F factor exists as a circular molecule within the cell but is transferred only as a single DNA strand which then replicates within the recipient cell. On average, an F^+ bacterial cell will contain one sex factor plasmid per cell chromosome, although the actual number is variable.

On occasions sex factor DNA may indulge in DNA exchange with part of the host cell DNA by recombination. This yields sex factors which, though often deficient in normal F^+ genes, do carry a few bacterial genes. Such factors are termed F prime (F′) factors, and it was obviously the transfer of such factors which led to the original discovery of bacterial conjugation by Lederberg and others. Actually such transfer of bacterial genes is a comparatively rare event in conjugation of normal F^+ and F^- strains, it is an interesting phenomenon in many ways analagous to bacterial transduction, (discussed on p. 68), in which a phage carries bacterial genes from one cell to another.

As we have stressed, in normal conjugation transfer of host genes is rare, but in 1958 Hayes and Luigi independently found certain strains of *E. coli* in which transfer of host cell genes was very common during conjugation. These strains they termed Hfr strains, meaning high frequency recombination strains, and indeed it was the subsequent analysis of such strains which led to our present knowledge of the mating process in bacteria and enabled gene order in *E. coli* to be mapped. It has now become clear that Hfr bacteria are those in which the sex factor has become integrated with the bacterial chromosome in the same way as phage lambda integrates as a prophage in lysogenic bacteria. An Hfr bacterium produces F pili just as does an F^+, but on conjugation a very different kind of transfer ensues. Rather than the sex factor itself passing over, the bacterial genome is transferred as a linear molecule. The genome appears to become linear because of a break at the site of insertion of the sex factor. Moreover a polarity is demonstrated by the chromosome in that the end next to the plasmid DNA is always transferred first and the integrated sex factor last. But, in reality, transfer of the whole chromosome is very rare and so passage of the sex factor in Hfr conjugation is also rare. Most frequently only a few bacterial genes are transferred and the chances of a gene transferring is directly correlated with its proximity to the leading free end of DNA. The bacterial DNA which is transferred is, like the plasmid itself during transfer, single stranded, but becomes double stranded again by replication after passage into the recipient cell.

When an F factor becomes integrated into an *E. coli* chromosome, it seems to do so at random and in either orientation. It follows that many different strains of Hfr bacteria will exist, each one transferring at high frequency marker genes close to the site of insertion, some clockwise and some anticlockwise round the chromosome. Analysis of the frequency of marker gene transfers has enabled the markers to be mapped and the circularity of the *E. coli* chromosome to be confirmed. Such mapping has been greatly facilitated by interrupted mating experiments in which DNA transfer from Hfr to F⁻ bacteria is artificially stopped by agitation in a high speed blender. The time of agitation after mixing of the strains can be closely correlated with the transfer of a particular marker gene.

Before we conclude our discussion of sex factor, one or two other points of interest deserve comment. The first is the remarkable phenomenon of plasmid *exclusion*. If F⁺ or Hfr strains are mixed with F⁺, chromosomal or sex factor transfer is rare, the sex factor already present apparently preventing DNA acceptance. The precise mechanism operative in exclusion is still not clear. A second phenomenon briefly alluded to is the strict control which *E. coli* has over the number of F plasmids per cell. Such control is said to be stringent, in comparison to the *relaxed* control found with some other bacterial plasmids. Stringent control normally implies that the non-integrated sex factor, itself a minicircle of DNA, is permitted to replicate only in time with the main bacterial DNA replication. Since the two pieces of DNA do not seem to be physically connected, it is presumed to result from strict cellular control over the initiation of DNA synthesis.

3.3.2 Colicin Plasmids

Many bacteria produce diffusable proteins which are lethal to certain other strains of bacteria, although not to the strain producing them. Such substances are termed bacteriocins: those produced by *E. coli*, are called, more specifically, colicins. The genes for these proteins are known to be carried on a plasmid and such Col. plasmids we will now discuss.

Col plasmids very neatly demonstrate one aspect of plasmid biology, namely, transmissibility. In certain Col plasmids, for example Col I, when a strain of *E. coli* which possesses Col I is mixed with a Col⁻ strain, transfer of Col I plasmid occurs. Such transfer is strictly analogous to sex factor transmission since again special pili are involved, these ones being distinct from sex pili and being termed I pili. But with other colicin plasmids, self transmission does not occur. Actually the transfer of these factors is possible in certain circumstances, in particular when sex factor is being transferred. Although the

F$^+$ and Col plasmids do not seem to be physically linked during transfer, the transfer of the normally non-transmissable Col factor is *mobilized* by the sex factor conjugation process.

We should also stress that transfer and spread of Col I plasmid is less infectious in a culture of bacteria than is sex factor. This seems to be because this plasmid is subject to control by repression and, in many cells which have harboured the plasmid for some time, expression of proteins coded on the Col I plasmid is repressed.

3.3.3   Drug Resistance Factors

In the late 1950's certain strains of *Shigella* bacteria, isolated in Japan from cases of human dysentery, were found to be simultaneously resistant to up to four separate antibiotic drugs. It soon became apparent that such multiple drug resistance could be transmitted from one bacterium to another as a single event, and even between bacteria of different species. The implications of this remarkable phenomenon for modern medicine are of course very great, but of more significance to us here is that the genes which determine such drug resistance are found to be borne on a bacterial plasmid, or R factor, and multiple drug resistance is explained by one R factor coming to carry more than one such gene at the same time.

R factors are of about the same general size as F and Col plasmids, being circular molecules with a molecular weight of about 7.0×10^7. Their effect in conferring resistance to antibiotics is attributable to their possessing genes coding for enzymes which specifically inactivate or degrade antibiotics. Some R factors may be quite numerous in the cells possessing them, in some cases up to 60 copies per cell being present. In other words, control of R plasmid replication may be *relaxed*, in comparison with the very *stringent* control of F factor replication. But this relaxed control is true of only a few resistance factors. Once again, R factors are transmissible plasmids and their presence leads to the formation of pili at the host cell surface, some resembling F pili, others I pili.

Although what has been stated about R factors is substantially correct, at the time of writing there is some indication that the genes for transmissible drug resistance are highly mobile within the cell, and may move easily between the plasmid DNA and the host chromosome. Pieces of DNA with this tendency to move their location or 'jump' have come to be known as transposons. It is as yet premature to guess how many aspects of plasmid integration into the host genome are attributable to these *insertion elements*, but one fascinating aspect of these transposable elements is their ability to cause effective mutations to host genes located close to the site of insertion. Such insertion elements seem to be a frequent type of sequence in the *E. coli* genome. Not only with R plasmids but with sex factor also, an insertion element sequence may occur on the plasmid and main genome which often accounts for plasmid integration. So we should understand these elements both as important factors in recombination and also as sequences which may profoundly influence the activity of adjacent genes in terms of both mutation and transcription.

3.3.4   Crown Gall Plasmid

For many years it has been known that a bacterium *Agrobacterium tumefaciens*, was responsible for inducing a common disease of plants, crown gall. The disease has some curious aspects, one being that the plant growth induced by the disease is neoplastic in character, so that crown gall has been looked on by some as a plant tumour. Yet more interesting is the observation that, once induced, the presence of the bacterium is not essential for persistence of the tumour and indeed that the disease, once induced by the bacteria, could be transmitted by tumour cells free of apparent bacteria.

All of this evidence led to a diligent search for a tumour-inducing principle (TIP) in crown gall, and especially in tumour inducing strains of the causative bacterium. This search has now been blessed with success since the tumorigenic strains of the bacterium have been found to carry a large circular DNA plasmid of about 1.10×10^8 daltons, while non-tumorigenic strains of the same bacterium lack the plasmid. Secondly, strains capable of inducing crown gall become ineffective when grown at $37°C$, a temperature also found to induce loss of the plasmid. The greatest success in the search for TIP of crown gall has come with the demonstration by both DNA and RNA hybridization studies that the crown gall cells of affected plants possess and express a segment of this plasmid. Interestingly, only part of the plasmid seems to be transferred to the plant cells, each cell being calculated to carry about 20 copies of a plasmid segment between 3×10^6 and 6×10^6 daltons mol. wt.

Crown gall can now be notched up as the first example of a disease of eukaryotes caused by a feat of natural genetic engineering, the transmission of a plasmid segment from bacterial donor to eukaryotic host cell.

3.3.5   Chimaeric Plasmids

These are plasmids which carry DNA from both bacterial and eukaryotic cells and are the result of artificial manipulation in the laboratory, most frequently involving the use of specific restriction endonuclease enzymes. This subject has been fully discussed in Chapter 4, having become a crucial technique in the investigation of eukaryotic gene sequences.

3.4   THE GENOMES OF MITOCHONDRIA AND CHLOROPLASTS

All eukaryotic cells contain mitochondria (except for one species of *Amoeba*) and all plant cells also contain chloroplasts or equivalent plastids. It has been known for many years that both of these organelles possess DNA, yet their presence in the cell is by no means optional, since the respiratory and photo-synthetic functions of eukaryotic cells entirely depend on their presence. We will not here expand on the many interesting aspects of biology and biochemistry concerned with these structures but it does seem appropriate to discuss some of their characteristics to the extent that they bear on the possible evolutionary relationship between prokaryotes and these organelles of eukaryotic cells. A plausible case can be advanced to support the notion that mitochondria are derived from endosymbiotic bacteria, and chloroplasts perhaps from blue-green algae, although the idea remains speculative. And there is some evidence that is difficult to reconcile with the theory. For example, ribosomal subunits from mitochondria and bacteria are not interchangeable, although chloroplast and bacterial subunits are. Secondly, the gene for the largest RNA is split in the yeast mitochondrion, and it is highly likely that so too are the genes for some of the cytochrome enzymes, yet no bacterial genes are yet known to have introns within them (see discussion on p. 166 and 253).

At the risk of seeming to favour the evolutionary connection between mitochondria and bacteria too strongly, it is perhaps appropriate to point out that endosymbiotic bacteria do indeed occur in a number of eukaryotic cells, none more remarkable than the kappa particles of *Paramecium*, now styled *Caedobacter taeniospiralis*.

Let us now return to examine the genomes of these organelles in more detail, and in particular the relationship between the nuclear genome and the DNA of the organelles. The mitochondrial genome consists of a covalently closed circle of double stranded DNA, of about 5 μm total length in higher animals but 25 μm in yeast and up to 30 μm in some plants. As with the circular DNA of plasmids and some phages, this DNA shows a number of superhelical turns or supertwists. There is evidence to support the idea that the mitochondria of higher organisms are all maternally derived and that the mitochondrion in the sperm is lost or destroyed after fertilization. For example, mules possess only horse type mitochondrial DNA while hinnies have donkey type (mules result from male donkey × female horse, hinnies from male horse × female donkey). Chloroplasts are maternally derived in some species, but in others they originate from both parents.

Not only are mitochondria presumed to be maternally derived (which implies a more than 50% genetic relationship to our mothers), but there is ample evidence that no copy of the mitochondrial genome occurs in the nuclear DNA. That is, mitochondrial genes are truly autonomous and do not simply represent extra 'spare' copies of nuclear genes. But it is equally certain that many, or indeed most mitochondrial proteins are coded in the nuclear genome (Table 3.4). The mitochondrial genome of animal cells is adequate for some 70 structural genes, but whether that number is actually present is not known. To date genes known to occur on the mitochondrial genome are as follows: 2 ribosomal RNA genes; about 20 transfer RNA genes; a number of genes coding for cytochrome enzymes 6 and cytochrome oxidase; some genes coding for drug resistance factors (these factors are probably enzymes involved in oxidative phosphorylation). Curiously enough the two mitochondrial rRNA genes are adjacent in *Xenopus* and man but widely spaced in yeast. Both DNA strands carry genes and indeed the tRNA genes are distributed some on one strand and some on another. Perhaps most curious of all is that, although mitochondrial DNA is replicated by a unique DNA polymerase, that enzyme is coded on the nuclear genome and made on cytoplasmic ribosomes. The ribosomal proteins of mitochondria, at one time thought to be exclusively coded on the nuclear genome, are now believed, in some organisms, to be coded and assembled within the mitochondria.

To sum up our short discussion of mitochondrial DNA, it is clear that this structure is highly dependent on nuclear genes for its structure and function, although many elaborate interractions are involved in what is essentially a state of genetic interdependence. If mitochondria began life as symbiotic bacteria they have certainly lost to, or delegated to, the host cell nucleus much of the genetic responsibility for their survival. Figure 3.28 illustrates a molecule of mitochondrial DNA in the act of replication and Fig. 3.29 a very provisional genetic map of yeast mitochondrial DNA.

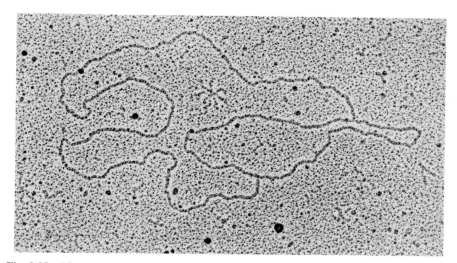

Fig. 3.28 Electron micrograph of a molecule of mitochondrial DNA isolated from rat liver cells. This molecule shows two replication forks (arrowed). Kindly supplied by Dr. D. Wolstenholme. (Wolstenholme D. R. *et al.* (1973) *Cold Spring Harbor Symposium*, **38**, 267–80.)

3.4.1 Recombination of Mitochondrial DNA in Yeast

Although the point is often made that our knowledge of the mitochondrial genome is greatly hampered by the lack of a recombination phenomenon,

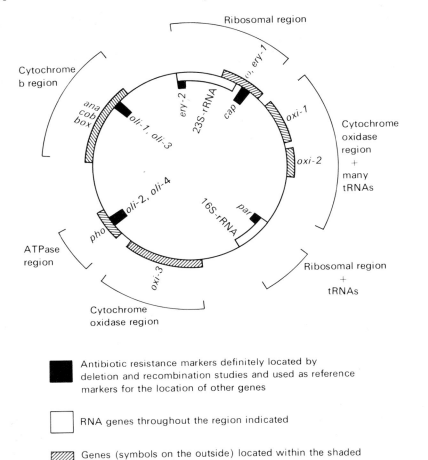

Fig. 3.29 Provisional genetic map of yeast mitochondrial DNA. (After various authors in Saccone C. & Kroon A. M. (1976) *The Genetic Function of Mitochondrical DNA.* North Holland, Amsterdam.) Gene symbols: *cap*—chloramphenicol resistance; *ery*—erythromycin-resistance; *oli*—oligomycin-resistance; *par*—paromomycin-resistance; *oxi*—cytochrome oxidase; *cob*—cytochrome b; *ana*—antimycin-A; *box*—cytochrome b and cytochrome oxidase; *pho*—oligomycin-sensitive ATPase; *ω*—polarity or sex factor. (From Beale C. & Knowles J. (1978) *Extranuclear Genetics.* Edward Arnold, London.)

Thomas and Wilkie discovered a recombination process in yeast mitochondria in 1968. The phenomenon was apparent in crosses between certain yeast strains carrying genes for chloramphenicol and erythromycin resistance on their mitochondrial genomes. Some cells resulting from such matings were sensitive to both drugs, and it seems impossible to explain this finding on grounds other than recombination between the mitochondrial DNAs of the two mated strains.

Table 3.4 Genetic control of mitochondrial constituents (provisional summary). (From Beale G. & Knowles J. (1978) *Extranuclear Genetics*. Edward Arnold, London.)

Constituent	Controlled by nuclear genome	Controlled by mitochondrial genome
1 Mitochondrial DNA-polynucleotide structure	—	√
2 Factors controlling replication, recombination, transcription of mitochondrial DNA	√	?
3 Protein synthesizing system in mitochondria:		
rRNA	—	√
ribosomal proteins	√	?
tRNA	√	√
4 Soluble enzymes	√	—
5 Other proteins:		
ATP-ase	√	√
Cytochrome oxidase	√	√
Cytochrome b	√	√
Other cytochromes	√	?

This recombination process has been found to display some unusual aspects, and, in particular, that the genetic exchange often seems to be polarized and non-equal. But the evidence is complex and a little confused so we need not discuss it further here.

3.4.2 Chloroplast DNA
Most of our discussion above has concerned mitochondria and their DNA. We will now examine the chloroplast in a little more detail. We must begin by stressing that chloroplast DNA is much larger than mitochondrial DNA, being at least eight times larger than the mitochondria DNA of mammals and therefore having enough to code for some hundreds of proteins. Genes known to occur on the chloroplast genome are akin to those of mitochondria i.e. probably 2 distinct rRNA genes and over 20 tRNA genes, although the evidence is less good than for mitochondria. A distinct chloroplast RNA polymerase is known, but the location of its gene is uncertain. A few other proteins are believed to be coded on the chloroplast genome, including some of the chloroplast ribosomal proteins. One particular protein which is known to be coded on the chloroplast genome and synthesized on chloroplast ribosomes is the main subunit of the fraction I protein which catalyses the first step in CO_2 fixation. But interestingly, this enzyme is a dimer, of which the other smaller subunit is encoded in the nuclear genome. This important protein, which accounts for some 40% of the soluble fraction of the plant chloroplasts, is then the result of a singular co-operation between nuclear and chloroplast genes.

In conclusion, it should also be pointed out that a process which parallels the observed recombination of yeast mitochondria has been recorded with the chloroplasts of the aquatic alga *Chlamydomonas* by Ruth Sager and others. So it is probable that our knowledge of the genetic map of chloroplast DNA may be enhanced by experiments with this useful organism.

3.5 FURTHER READING

Bachmann B. J., Low K. B. & Taylor A. L. (1976) Recalibrated linkage map of *E. Coli* K.12. *Bact. Rev.* **40**, 116–67.

Bainbridge B. W. (1980) *The Genetics of Microbes*. Blackie, Glasgow.

Beale G. & Knowles J. (1978) *Extranuclear Genetics*. Edward Arnold, London.

Birky C. W. (1978) Transmission genetics of mitochondria and chloroplasts. *Ann. Rev. Genet* **12**, 471–512.

Borst P. & Grivell L. A. (1978. The mitochondrial genome of yeast. *Cell* **15**, 705–23.

Broda P. (1979) *Plasmids*. Freeman & Co, Oxford.

Bukhari A. I., Shapiro J. & Adhya S. (1977) *DNA Insertion elements, plasmids and episomes*. Cold Spring Harbor Laboratory. New York.

Cohen S. N. (1976) Transposable genetic elements and plasmid evolution. *Nature* **263**, 731–8.

Dickson R. C., Abelson J., Barnes W. M. & Reynikoff W. S. (1975) Genetic regulation: the *Lac* control region. *Science* **187**, 27–35.

Doi R. H. (1977) Role of RNA polymerases in gene selection in prokaryotes. *Bact. Rev.* **41**, 568–94.

Fiers W. *et al.* (1978) Complete nucleotide sequence of SV40 DNA. *Nature* **273**, 113–20.

Flavell A. (1981) Did retroviruses evolve from transposable elements? *Nature* **289**, 10–11.

Gross H. J. *et al.* (1978) Nucleotide sequence and secondary structure of potato spindle tuber viroid. *Nature* **273**, 203–8.

Helinski D. R. (1978) Plasmids as vehicles for gene cloning. *Trends in Biochem. Sci.* **4**, 10–14.

Kelly T. J. & Nathans D. (1971) The genome of Simian virus 40. *Adr. Virus Res.* **21**, 86.

Lewin B. (1974) *Gene Expression—1. Bacterial Genomes*. John Wiley & Sons, Chichester.

Lewin B. (1977) *Gene Expression—3. Plasmids & Phages*. John Wiley & Sons, Chichester.

Lewin B. (1976) DNA sequences coding for more than one protein. *Nature* **264**, 11–12.

Lippincott J. A. (1977) Molecular basis of plant tumour induction—News & Views. *Nature* **269**, 465–6.

Meynell G. G. (1972) *Bacterial Plasmids*. Macmillan, London.

Neren P. & Saedler H. (1977) Transposable genetic elements as agents of gene instability and chromosomal rearrangements. *Nature* **268**, 109–15.

Pennington T. H. & Ritchie D. A. (1975) *Molecular Virology*. London. Chapman & Hall.

Saedler H. (1977) The role of IS-elements in the evolution of *E. Coli* chromosome and some of its plasmids. In Bradbury & Javaherian (eds.) *The organisation and expression of the eukaryotic genome*. Academic Press, London and New York.

Sanger G. M. *et al.* (1977) Neucleotide sequence of bacteriophase ϕX174 DNA. *Nature* **265**, 687–95.

Sambrook J. (1977) Adenovirus amayes at Cold Spring Harbor. *Nature* **268**, 101–4.

Schwesinger M. D. (1977) Addition recombination in bacteria. *Bact. Rev.* **41**, 872–902.

Szekely M. (1978) Triple overlapping genes. *Nature* **272**, 492.

Wood W. B. & Revel H. R. (1976) The genome of bacteriophage T4. *Bact. Rev.* **40**, 847–68.

Chapter 4
Sequence and Gene Arrangements in DNA

The information required to construct and maintain a cell or, indeed a whole organism, is encoded in the sequence of nucleotide pairs along its DNA molecules. Consequently, the determination of this sequence is fundamental to understanding molecular biology. The first substantial nucleic acid sequence was published in 1965 and earned Professor Holley a Nobel prize. Since then enormous improvements in the technology of sequence determination (section 4.11) have resulted in the determination of sequences of over 5000 nucleotides, for example, the entire DNA of the small bacteriophage called ϕX174 and the DNA of the eukaryote virus sV40. A human cell contains about 1×10^9 nucleotide pairs so the complete sequence is going to be hard to determine and difficult to understand once it has been determined. For many purposes, it is more useful to know how large sections of DNA (e.g. genes) are organized with respect to each other and with respect to other types of DNA sequences, e.g. control regions. As we shall see, this kind of information can be obtained using techniques such as nucleic acid reassociation, hybridization, restriction site mapping and cloning, but first let us examine the so-called 'C-value paradox'.

4.1 DNA CONTENT

The C-value is the amount of DNA contained in one haploid cell. This varies from species to species but within one organism the 'C-value' is approximately constant for most cells. For simple organisms the C-value increases as the sophistication of the organism increases, from a few thousand nucleotides in a simple virus to several million in the bacterium *Escherichia coli* and tens of millions in lower eukaryotes like yeast. Simple organisms use their DNA very efficiently and almost every nucleotide is involved in at least one coding sequence or in a short control sequence. However, if we look at some other lower eukaryotes and at higher eukaryotes we find that although none has a small C-value there is a wide variation in C-value even among closely related species. For example, in amphibians the haploid genome size varies between 9×10^8 base pairs and 8×10^{10} base pairs, a range of nearly one hundred-fold. This is a very surprising fact if we regard DNA as just 'information'. Why should closely related species possess very different C-values? Why do most eukaryotes possess so much more DNA per cell than prokaryotes? These questions are known as the C-value paradox.

This question is still open but it has been narrowed a little by calculations and measurements of how much 'information' we would expect to find in the DNA of a higher organism.

The calculations assume that all genes are subject to mutation which are usually harmful. If an organism has a very large number of genes then some genes in each individual will undergo harmful mutations and the individual or its progeny will die prematurely. The average number of mutations per gene per year can be measured in suitable species like the fly, *Drosophila*, and the number of genes that can be present before the overall mutation rate becomes lethal can then be estimated. The details of the estimation are controversial but most scientists accept that a human genome can only contain 5000–50 000 different genes.

The measurements involve polytene chromosomes (see Chapter 6), an entirely different approach, where genes can be visualized (see 6.22) on giant chromosomes. Again there is some controversy but there are probably about 5000 different genes in the DNA of one cell containing giant chromosomes.

Five thousand to fifty thousand genes account for up to 10% of the DNA of a human cell. This is called coding DNA; the remainder is called non-coding DNA. It seems unlikely that the C-value paradox reflects differences in the amount of coding DNA in related organisms. It is more likely that C-value variations reflect variations in the amount of non-coding DNA. A large amount of non-coding DNA is probably a feature of all higher eukaryote cells. Another surprising feature of eukaryote DNA has been revealed by DNA reassociation experiments.

4.2 DNA REASSOCIATION

The two strands of the DNA double helix can be separated by a number of methods, including raising the temperature of a DNA solution or dissolving the DNA in a denaturing agent like formamide. This is a useful reaction and we will return to it in Section 4.5. Now, however, consider the reverse reaction, the reassociation of the single strands of DNA to form the original double strand. If this works correctly the new double helix will be fully base paired and such a structure can only be formed by the reassociation of two strands with complementary sequences. Consequently, the reaction is fairly slow, because each DNA single strand must move round, in solution, constantly colliding with other DNA single strands until it finds a complementary strand and they collide with their complementary bases opposite one another. Clearly, the more concentrated the DNA the more frequent the collisions will be, and hence the faster the reaction proceeds. This means that the amount of double stranded DNA formed depends on the time the reaction has been going, t, and on the original concentration of complementary single strands, c_0. The product of these two quantities, $c_0 t$, is called 'cot'. For the ideal case of a monodisperse solution of the original, non-repetitive, un-nicked, double strands the rate of reassociation after complete disassociation is governed by a normal second order rate law,

$$\frac{dc}{dt} = -kc^2$$

where k is a constant and c is the concentration of dissociated strands.

The constant, k, depends on the reaction conditions particularly temperature, pH, ionic strength and the presence of denaturing agents such as formamide. The outline plan of a measurement of DNA reassociation is shown in Fig. 4.1. The DNA used needs to be free from RNA, which would compete for complementary binding sites (Section 4.8), and protein which might affect the stability of the single and/or double stranded forms of DNA. This state of purity is usually achieved by treating the DNA with ribonuclease and

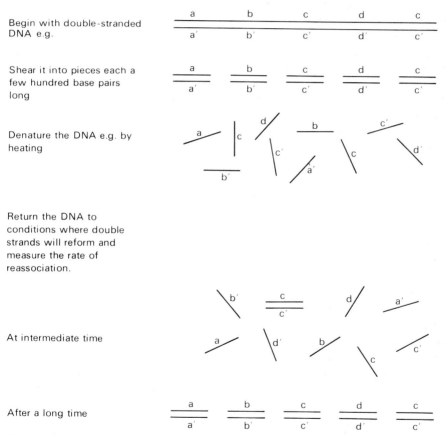

Begin with double-stranded DNA e.g.

Shear it into pieces each a few hundred base pairs long

Denature the DNA e.g. by heating

Return the DNA to conditions where double strands will reform and measure the rate of reassociation.

At intermediate time

After a long time

Fig. 4.1 Measurement of DNA reassociation. a, b, c, d are DNA sequences and a′, b′, c′ and d′ are their complementary sequences. Notice that cc′ is repeated and so the probability of reassociation is higher for these sequences.

protease followed by extraction of any remaining protein material with phenol and precipitation of the DNA with ethanol. The DNA is then sheared to the length, usually a few hundred base pairs, that gives stable double helices but allows individual sequences to reassociate independently of one another.

The DNA concentration, c_0, is measured, usually from its ultraviolet absorbance, and expressed as moles nucleotide per litre, mol. 1^{-1}. The DNA is denatured, usually by a combination of formamide and high temperature, and then incubated at a lower temperature where the single strands will reassociate with complementary base pairing. A variety of initial concentrations c_0 and incubation times, t is used.

The progress of the reaction can be measured directly from the ultraviolet absorbance because double stranded DNA has a lower absorbance than single stranded DNA at wavelengths near 260 nm. This reduced absorbance results from the 'stacking' of the bases that occurs in double stranded DNA. Hence,

$$\text{fraction reassociated} = \frac{A(s) - A(r)}{A(s) - A(d)}$$

where $A(s)$, $A(r)$ and $A(d)$ are the absorbance at 260 nm of single stranded DNA, reassociated DNA and double stranded DNA respectively. When the DNA is only available in small amounts, or very low concentrations have been used, the experiments are carried out with radioactive DNA of known specific activity. The incubation is then stopped by chilling the solution and the fraction reassociated is measured in one of two ways: (1) a nuclease (e.g. S1 nuclease) that specifically degrades single stranded DNA is added and the solution incubated. The undegraded, double stranded, DNA is precipitated and collected on a filter which is washed and placed in a counter for radioactivity determination. Then, fraction reassociated = fraction resistant to S1 nuclease; (2) the DNA solution is passed through a hydroxyapatite (HAP) column in 0.12 M phosphate buffer. Only double stranded DNA binds to the column under these conditions so the fraction reassociated = fraction of DNA bound to HAP.

Together there are three methods for determining the size of the reassociated fraction. The hydroxyapatite method gives different results from the other two methods because the newly formed DNA double strands will usually have single-stranded regions at one or both ends, since the two components are unlikely to have been broken at the same place originally. These ends are counted as reassociated DNA by the hydroxyapatite method but not by the other two methods, so the hydroxyapatite method gives a larger estimate of the reassociated fraction. When this is taken into account the three methods are in good agreement.

In the case of a simple viral DNA, with about 5000 base pairs of DNA per virus particle, the reaction is half completed after a $c_0 t$ value of about 0.01 mol s 1^{-1}. With a bacterial DNA containing about 5×10^6 base pairs of DNA per cell, at the same total DNA concentration as the virus DNA, the concentration of complementary strands is 1000 times less. Hence the reassociation of the bacterial DNA will require a $c_0 t$ value of 10 mol s l^{-1} to reach half completion. This value, $c_0 t$ for 50% reassociation, is called $c_0 t_{\frac{1}{2}}$ and it is a measure of the *complexity* of the DNA. In the above example, the $c_0 t_{\frac{1}{2}}$ of 0.01 mol s l^{-1} for the viral DNA corresponds to a complexity of 5000 b.p. or 5 kb, and the $c_0 t_{\frac{1}{2}}$ of 10 mol s l^{-1} for the bacterial DNA corresponds to a complexity of 5000 kb. The complexity of DNA is the length, in base pairs, before the sequence begins to repeat itself. It can be found from a knowledge of the actual sequence or by measuring the $c_0 t_{\frac{1}{2}}$.

Figure 4.2 shows the time course of a reassociation reaction first as 'fraction reassociated' plotted against time (Fig. 4.2(a)) and then as 'fraction reassociated' plotted against $c_0 t$ on a logarithmic scale (Fig. 4.2(b)). The logarithmic scale is used to expand the scale at short times (low $c_0 t$) and compress it at long times (high $c_0 t$) so that one scale can be used for widely different DNAs. It is also easy to measure $c_0 t_{\frac{1}{2}}$ using the logarithmic scale.

Higher eukaryotes have several hundred times as much DNA per cell as bacteria so the $c_0 t_{\frac{1}{2}}$ should be several thousand mol s l^{-1}. These $c_0 t$ values can be measured although they are difficult to reach because of the very high DNA concentration and long incubation required. However, a substantial proportion of any given higher eukaryote DNA reassociates at much lower $c_0 t$ values as shown in Fig. 4.2(b) for human DNA. Figure 4.2(b) shows that eukaryote DNA can be divided by rate of reassociation, into four classes called kinetic classes or sequence classes: slowly reassociating; intermediate; rapidly reassociating; zero time reassociating.

The slowly reassociating fraction behaves as expected for a DNA of high complexity and is referred to as *unique* or *single copy DNA*. The next two fractions behave as if they had a lower complexity. In other words each sequence must occur many times in the DNA of one cell, so increasing the effective concentration of complementary sequences and hence the rate of reassociation. The intermediate fraction is known as *intermediate repetitive DNA* and the rapidly reassociating fraction as *simple sequence DNA*. The zero time reassociating fraction represents *inverted repetitive sequences* (see section 4.5.1).

In human DNA, which is typical of higher eukaryote DNA, about 51% of the DNA sequence is unique, about 13% is intermediate repetitive and about 22% is highly repetitive or simple sequence DNA. Of the remainder 6–9% reassociates too fast to measure (zero time) and about 5% fails to reassociate. This last 5% is not interpreted as single stranded DNA regions but as experimental error due to factors such as occasional mismatching, and degradation to sequences too short to reassociate. These values, as in all reassociation experiments, are simplifications of the highly complex sequence organization in eukaryotes but they are useful in giving an overall picture and showing the presence of repetitive sequences. In particular, the amount of DNA found in the intermediate repetitive class depends on the reassociation reaction conditions in that conditions which allow some mismatching to occur give rise to a larger fraction of DNA in the intermediate repetitive class whereas stringent

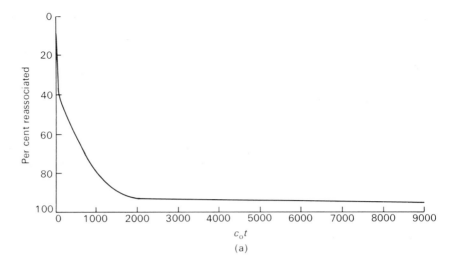

(a)

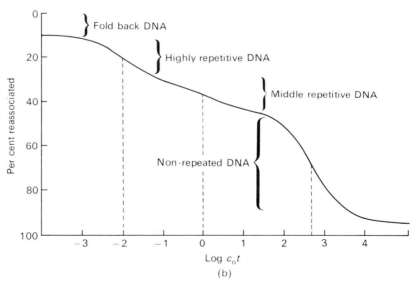

(b)

Fig. 4.2 The reassociation profile of highly sheared human DNA. The self-reassociation profile in 0.24 M sodium phosphate at 65 C of 0.6 kb strands of human DNA must be represented by several kinetic classes. The curve is made from 4 components: 51 % with $c_0 t_{\frac{1}{2}} = 500$ (non-repeated DNA); 12 % with $c_0 t_{\frac{1}{2}} = 1.0$ (middle repetitive DNA); 22 % with $c_0 t_{\frac{1}{2}} = 0.01$ (highly repetitive DNA) and 9 % 'zero time' reassociating (fold-back DNA). Four kinetic classes is the minimum number consistent with the data but other repetition frequencies may be present.

(a) $c_0 t$ on a linear scale (approximate values).

(b) $c_0 t$ on a log scale.

(Adapted from Schmid C. W. & Deininger P. L. (1975) *Cell* **6**, 347.)

conditions where little mismatching occurs give rise to a larger fraction of DNA in the unique sequence class. We will consider these four different DNA fractions further, but first let us ask what other evidence there is for the existence of repetitive DNA sequences in eukaryotes.

4.3 DNA SATELLITES

The most direct confirmatory evidence for repetitive DNA comes from actual base sequence studies of simple sequence DNA where the highly repetitious nature of the sequences can be seen, for example, one guinea pig satellite is more than half composed of the sequence

$$
\begin{array}{cccccc}
T & T & A & G & G & G \\
A & A & T & C & C & C
\end{array}
$$

and a satellite of the hermit crab is

$$
\begin{array}{cccc}
T & A & G & G \\
A & T & C & C
\end{array}
$$

repeated many times. Evidence for one form of intermediate repetitive DNA comes from electron microscope pictures of transcribing rRNA genes (Fig. 4.3) where the repetition can be seen directly. More general evidence for the existence of fractions of DNA with repeated sequences comes from density gradient centrifugation. When a solution of caesium chloride (CsCl) is centrifuged, the dense Cs^+ ions tend to move to the outside of the tube or centrifuge cell and thus a density gradient is formed. DNA dissolved in the CsCl solution will, on centrifugation, form a band at the point in the density gradient where the DNA density equals the density of the CsCl solution. The position of the DNA band can be found by fractionating the gradient or by scanning it directly in an analytical ultracentrifuge. The density of DNA and hence its position in the gradient depends on its base composition. The resolution can be enhanced by using ions (e.g. Ag^+) or antibiotics (e.g. actinomycin, netropsin) that bind to specific DNA sequences. When eukaryote DNAs are examined there is usually a main band and one or more minor bands called satellites. The main band represents a mixture of DNA molecules derived from the cellular DNA by random breakage during preparation. These DNA molecules have different base compositions due to their different sequences and so they band in slightly different positions in the density gradient. The presence of an extremely large number of different DNA molecules gives rise to very many overlapping bands that combine to produce a single broad band, the main band. However, if a particular sequence is repeated many times within the cell there will be a substantial proportion of the DNA molecules with this sequence and hence a particularly large homogeneous band at the appropriate density for this sequence. If this band is near the centre of the main band it will not be easily

Fig. 4.3 Electron micrograph of nucleolar genes in the process of transcription, with Xmas trees of ribosomal RNA apparent. Isolated from the oocyte of the newt *Triturus viridescens* (× 14 000). (Photograph kindly supplied by Prof. O. L. Miller.)

seen and is known as a cryptic satellite; if the band is away from the centre of the main band it will show as a separate satellite band.

The satellite DNA bands are due to DNA sequences that occur many times in the genome. The simple sequence DNA fraction can usually be isolated as one or more satellites and, as we shall see later, some gene-containing sequences are reiterated several hundred times, which enables them to be separated as satellites. Figure 4.4 shows examples of these satellites. There is

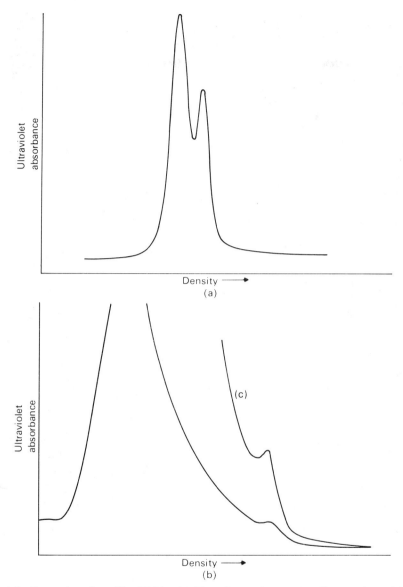

Fig. 4.4 Examples of satellite DNAs. (a) The reiterated genes for ribosomal RNA in *Physarum*. The diagram shows the optical density of DNA in a density gradient of CsCl. The left hand peak is bulk or main band DNA. reduced in amount in this case by partial purification of the satellite; the right hand peak is the satellite.

(b) & (c) A simple sequence satellite from cat DNA. In this case the satellite has not been purified but the resolution of the density gradient has been increased by the use of netropsin. The satellite is the small peak on the right. shown 4 × expanded in (c). The major peak is the bulk or main band DNA.

some evidence for the existence of other satellites, possibly formed from other types of intermediate repetitive DNA. CsCl gradients provide a good method for separating satellite DNAs and have been used to prepare simple sequence DNA, genes for ribosomal RNA and genes for histones.

4.4 INTERSPERSION OF MIDDLE REPETITIVE AND SINGLE COPY DNA

As we have seen, in most organisms, the bulk of the DNA in one haploid chromosome set has a nucleotide sequence that occurs either once (single copy) or about 100 times (middle repetitive). We can ask whether these two types of sequence occur either: (1) separately from one another in a small number of very long sequences that are exclusively single copy or middle repetitive; or (2) mixed at random with a very large range of sizes of the single copy sequences, and a very large range of uncorrelated sizes of middle repetitive sequences; or (3) mixed in an ordered fashion, with restricted or correlated ranges of sizes for the single copy sequences and the middle repetitive sequences. The way in which the sequences are mixed up is called the *interspersion pattern* and a part of the DNA that contains only one type of sequence is called a *block* of single copy DNA or a block of middle repetitive DNA (and similarly for the other two DNA sequence classes, simple sequences and inverted repeat sequences).

The interspersion pattern can be determined by studying the products of DNA reassociation reactions taken to an intermediate $c_0 t$ value such that middle repetitive sequences are almost entirely reassociated while almost none of the single copy DNA is reassociated. This gives rise to reassociated DNA molecules which can be studied by electron microscopy, hydroxyapatite chromatography, digestion with S1 nuclease and other methods.

Electron microscopy shows that most of the reassociated molecules contain both double strand regions (due to reassociated middle repetitive sequences) and single strand regions (due to single copy sequences adjacent to middle repetitive sequences). One such molecule is shown in Fig. 4.5(a). In this case the length of the block of middle repetitive sequence can be measured directly. It is 8.3 kb. A very large number of such molecules, chosen at random, can be measured and the experiment repeated with different size DNA fragments in the original reassociation reaction to give different sized reassociated DNA molecules. The results treated statistically then provide an overall picture of the interspersion pattern of the middle repetitive and single copy DNA sequences.

Electron microscopy gives a very clear picture of the structure of individual reassociated DNA molecules. However, it is hard to measure enough molecules to obtain an accurate overall picture of the interspersion pattern. The pattern will also be affected if the DNA molecules that are measured are not truly

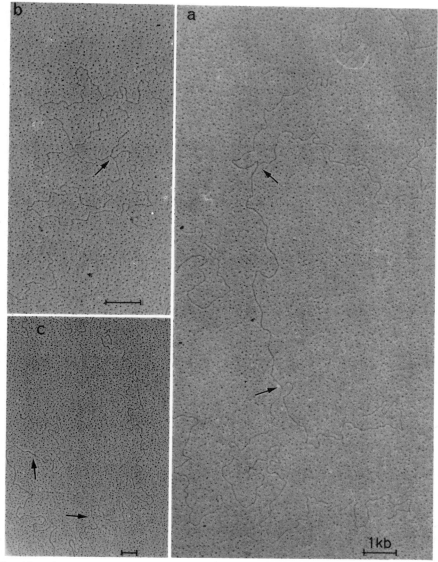

Fig. 4.5   Electron Microscopy of reassociated DNA. (a) The region of DNA between the arrows is reassociated, the 'tails' have not reassociated and so are assumed to be non-complementary.

(b) Similar to (a) but the length of complementary DNA which has reassociated is very short.

(c) The region between the arrows is reassociated but there are only two 'tails' suggesting the DNAs have reassociated at their ends.

(Photograph kindly supplied by Dr N. Davidson, from Manning J. E., Schmid C. W. & Davidson N. (1975) *Cell* **4**, 145.)

typical of the main population. These difficulties can be overcome by using biochemical methods, for example hydroxyapatite chromatography and S1 nuclease digestion. S1 nuclease degrades single stranded DNA but not double stranded DNA. Hence S1 nuclease can be used to remove the single strand 'tails' from reassociated DNA molecules leaving the double strand region intact. The lengths of the double strands can then be measured and the results give the lengths of the blocks of middle repetitive sequences.

The most generally used approach relies on hydroxyapatite chromatography. A series of reassociation reactions is performed, each consisting of a large number of unlabelled DNA fragments of a fixed length (usually 0.45 kb) and a small amount of labelled DNA fragments. The labelled DNA fragments are of different lengths in the different reassociation reactions. The fragments are denatured and allowed to reassociate to $c_0 t$ value such that the middle repetitive sequences form double strands but the single copy sequences do not, as before. Each reaction mixture is then passed through a hydroxyapatite column under conditions where reassociated DNA molecules containing significant double strand regions are retained and free single strand DNA molecules pass through. The percentage of the labelled DNA remaining bound to the hydroxyapatite column is measured. The results for *Xenopus* DNA are shown in Fig. 4.6. The shape of the curve is analysed in terms of the curve that would be expected for the simplest ordered interspersion pattern, namely that all the blocks of middle repetitive sequences have the same length l_R and all the blocks of single copy sequence have the same length l_S. The fixed length (0.45 kb) unlabelled fragments are in vast excess so we can assume that every labelled fragment containing a middle repetitive sequence will reassociate in each reaction, independent of the fragment length. Therefore, at very short fragment lengths where the 'tails' of single copy DNA are negligibly short, the percentage of labelled DNA forming reassociated molecules is equal to the total percentage of middle repetitive DNA in the labelled fragments. Hence, the intercept on the abscissa gives the percentage of the genome that is in middle repetitive sequences; let's call it P_R. The results agree with those determined by kinetic analysis ($c_0 t$ curves). As the labelled fragment length is increased some 'tails' of single copy sequence appear in the reassociated molecules and so the total percentage of labelled DNA in reassociated molecules increases. This continues until the fragment length becomes greater than l_s, the length of the blocks of single copy DNA. Then every block of single copy DNA has at least some middle repetitive DNA attached and so all the labelled DNA appears in reassociated molecules. The single copy block length l_s is then given by the minimum fragment length at which all the labelled DNA is bound. We can also say that the percentage of interspersed single copy DNA P_S is equal to the additional percentage of the DNA that is bound as the fragment length is increased. If we take a fixed amount of DNA, say one genome, then the number of blocks of single copy DNA is the total amount of interspersed single copy DNA

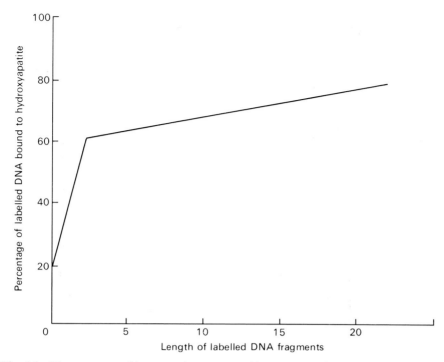

Fig. 4.6 Measurement of interspersion pattern of human DNA by reassociation. The curve shown is consistent with the following model. Tandemly repeated sequences occupy 6% of the genome; 0.6 kb long repeats interspersed with 2.25 kb long single copy sequences occupy 52% of the genome; and the remainder of the genome consists of very long single copy sequences interspersed with the 0.6 kb repeats. The precise numbers are not very well defined by this analysis but it is qualitatively correct. (Adapted from Schmid C. W. & Deininger P. L. (1975) *Cell* **6**, 347.)

divided by the amount of DNA in one block, which is proportional to P_s/l_s. There must be the same number (give or take one) of blocks of middle repetitive sequences, therefore

$$\frac{P_R}{l_R} = \frac{P_S}{l_s}.$$

Since we already know P_R from the intercept on the abscissa and P_S and l_s from the point where the curve stops rising we can find l_r by solving this equation:

$$l_R = \frac{P_R}{P_S} l_s.$$

We can also find l_R graphically because it turns out to be the numerical value of the intercept on the ordinate that is obtained by continuing the curve downwards in a straight line. The expected result for this simple model is shown in

Fig. 4.7 together with an indication of the types of reassociated DNA molecules present at various fragment lengths.

Comparison of Figs. 4.6 and 4.7 shows that the results are somewhat more complex than these given by the simple model. However, by an extension of the analysis to allow for some spread in the block lengths and for two basic interspersion patterns important conclusions can be drawn. For example in the case of *Xenopus* half of the DNA consists of a short period interspersion pattern, with single copy sequences of about 1 kb length spaced with middle repetitive sequences of about 0.3 kb length. Most of the remaining DNA is also interspersed but the lengths of single copy blocks are greater, over 4 kb. This is shown in Fig. 4.8a. It is called the *Xenopus pattern*.

Figure 4.8(b) also shows that *Drosophila* DNA has a different interspersion pattern. The single copy and middle repetitive sequences are present in the usual proportions in this DNA and they are interspersed with one another.

However, the lengths of the blocks of single copy sequence are very large, more than 13 kb and the lengths of the blocks of middle repetitive DNA are

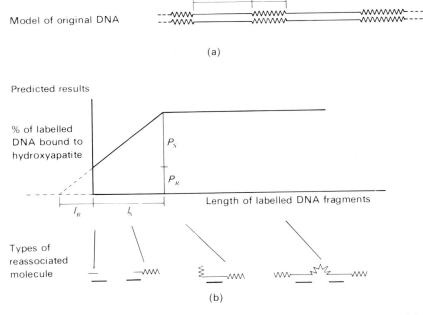

(a)

(b)

Fig. 4.7   Analysis of simplified interspersion measurement. (a) This simplified model of DNA shows single copy sequences of a fixed length l_S interspersed with repetitive sequences of fixed length l_R.

(b) Predicted results for the DNA in (a)

P_S and P_R are the percentages of DNA in interspersed single copy sequences and in repetitive sequences respectively. l_S and l_R are the lengths of the regions of single copy sequence and of the regions of repetitive sequences, as in (a) above.

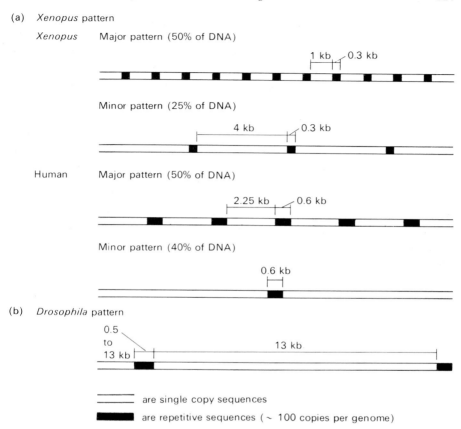

Fig. 4.8 Interspersion patterns. The arrangement of repetitive and single copy DNA sequences.

variable in the range 0.5–13 kb with a number average of 5.6 kb. This is known as the *Drosophila pattern*.

Subsequent studies on several other organisms have shown that the *Xenopus* pattern is very widespread among higher eukaryotes. Small differences, for example between *Xenopus* and the sea urchin *Stronglyocentrotus*, have been observed but they may be within experimental error. The substantial anomaly observed with *Drosophila* DNA (and that of some other insects) remains a mystery and is a factor that must be born in mind when it is desired to extrapolate other data on genome organization from *Drosophila* to mammals or other higher eukaryotes.

Finally, on the subject of interspersion, we can ask whether the structural gene sequences occur within the interspersed single copy sequences. Since the structural gene sequences code for proteins, by definition, they must be present in mRNA and so mRNA isolated from polysomes can be used as a probe for structural gene sequences. It is not a perfect probe since mRNA contains non-translated sequences and there may be structural gene sequences that are absent from mRNA populations as isolated from particular cells.

The interspersed single copy sequences (also called repeat-contiguous single copy sequences since they occur contiguously with repetitive sequences) are isolated by fragmenting DNA to a length of about 1.8 kb, then denaturing it and renaturing to a $c_0 t$ value such that middle repetitive sequences reassociate but single copy sequences do not. The reassociated molecules are isolated by hydroxyapatite chromatography and purified by one more cycle of dissociation, reassociation and hydroxyapatite chromatography. The reassociated molecules are then sheared to about 0.45 kb and dissociated and reassociated once more, but this time the single strand, non-reassociated, molecules are kept. These are single copy sequences because they did not reassociate the last time round. However, they must have been contiguous with a repetitive sequence in the 1.8 kb fragments in order to have survived the first two reassociation reactions. Hence these single strand molecules represent the interspersed single copy sequence. The method is summarized in Fig. 4.9.

In one particular experiment with sea urchin DNA interspersed single copy sequences representing one third of the total single copy sequence were isolated. However, hybridization with mRNA from sea urchin gastrulae showed that 80% of the mRNA sequences that hybridized to total DNA also hybridized to interspersed single copy DNA. The result is interpreted to mean that at least 80% of the structural gene sequences expressed in the gastrula occur adjacent to repetitive sequences in the chromosome DNA. The recent discovery of discontinuities in some structural gene sequences means that the interspersed repetitive sequences may occur within the structural gene sequence as well as at its ends.

A possible implication of the short period interspersion pattern (*Xenopus* pattern) is that the interspersed repetitive sequences are involved in the control of transcription of the structural genes present in the interspersed single copy DNA (see discussion in Chapter 7). Confirmation of this idea will have to wait for a much more detailed understanding of control of transcription in eukaryotes. However, it is worth pointing out here that there are many cases where a structural gene is distinguished from contiguous 'spacer' DNA sequences by a different overall base composition. This was first observed for the ribosomal genes in *Xenopus* where the thermal denaturation profile of the DNA containing the sequences for rRNA shows two separate transitions, one (A–T-rich) corresponding to spacer DNA and one (G–C-rich) corresponding to the rRNA sequences themselves. A similar distinction is found in *Xenopus* 5S DNA, *Xenopus* and yeast tRNA genes and sea urchin histone genes. It is not yet known if such a distinction also occurs with single copy genes. However, it is not universal for repetitive genes, either, since the genes for ribosomal RNA in *Physarum* show only a very small distinction of base composition between

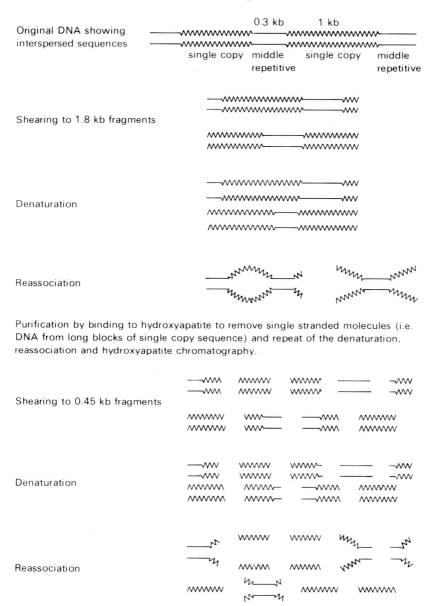

Shearing to 1.8 kb fragments

Denaturation

Reassociation

Purification by binding to hydroxyapatite to remove single stranded molecules (i.e. DNA from long blocks of single copy sequence) and repeat of the denaturation, reassociation and hydroxyapatite chromatography.

Shearing to 0.45 kb fragments

Denaturation

Reassociation

Purification of single strands by passing over hydroxyapatite. These are interspersed single copy sequences.

Fig. 4.9 Preparation of interspersed single copy DNA.

coding DNA and 'spacer' non-coding DNA. Nevertheless, the correlation of base composition and function of these DNA sequences is sufficiently common and striking to imply that it probably has some importance in the structure or expression of the DNA.

4.5 DNA CLASSES

Experiments such as those we have discussed reveal the existence of the four broad classes of DNA sequences: inverted repeats (zero time reassociating); simple sequence (highly repetitive); middle repetitive; and single copy (Fig. 4.10). In general, prokaryotes contain almost exclusively single copy DNA whereas eukaryotes contain all four classes of DNA. The proportion of DNA in each class varies from organism to organism. The measured proportions can also vary with the stringency of the reassociation conditions. Under stringent conditions very little mismatching of bases is allowed while under less stringent conditions mismatching, in the form of looped out or non-paired bases (Fig. 4.5), occurs and so a higher proportion of repetitive DNA is scored. DNA reassociated under stringent conditions has a thermal denaturation temperature, T_m, within $1°$ of native DNA of the same base composition whereas DNA reassociated under less stringent conditions has a lower T_m. This is used as a test for good reassociation.

The importance of molecular weight of DNA in reassociation reactions was discussed in the previous section (section 4.4).

4.5.1 Inverted repeats

Some 6 to 9% of human DNA occurs as short inverted repeats, also called palindromes. When the DNA is denatured and then allowed to renature each single strand will snap back on itself to form a double stranded hairpin structure (Fig. 4.10). Short inverted repeats would occur by chance in a random nucleotide sequence. For example, a class of DNA degrading enzymes called restriction nucleases (see section 4.6) recognizes short specific sequences that are inverted repeats. The sequence GAATTC which is recognized by *Eco* RI restriction nuclease would be expected to occur by chance approximately once every 3000 nucleotides and this frequency of occurrence is usually observed in DNA. However, sequences that are long enough to form stable hairpin structures are very unlikely to occur by chance in DNA of the complexity observed in eukaryotes and so inverted repeat sequences must be regarded as a separate class of sequences.

Inverted repeat sequences have been isolated from DNA by carrying out a DNA reassociation experiment at very low $c_0 t$, typically 10^{-6} mol s (called zero time reassociation) and isolating the double stranded DNA on hydroxyapatite.

The repeated sequences were then purified by digesting single stranded ends

1. Inverted repeat sequences (zero time reassociating)

Native structure (note
inverted repeat of a/a')

| 5' | a | b | c | a' | d | 3' |
| 3' | a' | b' | c' | a | d' | 5' |

Reassociated, hairpin,
structures

Palindrome

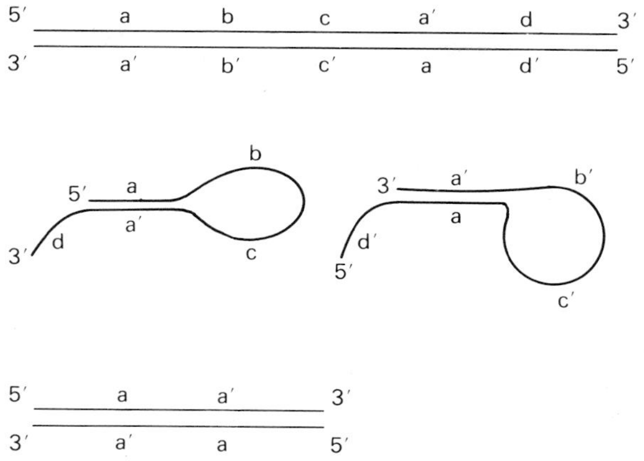

| 5' | a | a' | 3' |
| 3' | a' | a | 5' |

2. Simple sequences (rapidly reassociating) e.g. two hermit crab satellites

(i)   ...C C T A C C T A C C T A C C T A C C T A...
        ...G G A T G G A T G G A T G G A T G G A T...
(Note the repeat of CCTA · GGAT)

(ii)   ...C T G C A C T C A G C A G C A G C A G C A G C T G C A C T C A G...
        ...G A C G T G A G T C G T C G T C G T C G T C G A C G T G A G T C...

[Here the repeat is of the form CTGCACT(CAG)$_n$. The number of (CAG) repeats in a
row varies from 3 to 12]

3. Intermediate repetitive sequences (intermediate reassociating)

Tandem repeats e.g.
histone genes

| a | a | a | a | a | a |
| a' | a' | a' | a' | a' | a' |

Interspersed repeats e.g.
control sequences

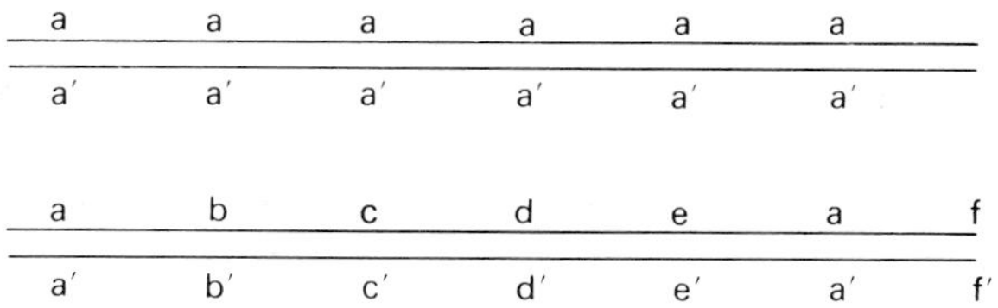

| a | b | c | d | e | a | f |
| a' | b' | c' | d' | e' | a' | f' |

4. Single copy sequences (slowly reassociating)

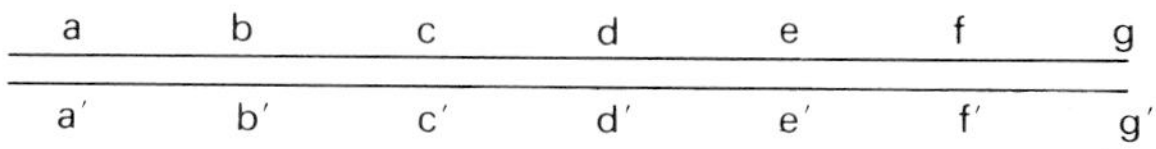

| a | b | c | d | e | f | g |
| a' | b' | c' | d' | e' | f' | g' |

Fig. 4.10 DNA sequence classes (kinetic classes). The letters a, b, c . . . represent
specific, different, DNA sequences and the letters a', b', c', ... represent the
corresponding complementary sequences.

and loops with S1 nuclease. The repeated sequences had to be at least 50 base pairs long to survive the isolation. In an experiment starting with DNA 15 kb long 26% bound to hydroxyapatite after zero time reassociation but only 6% of the original DNA also survived the S1 nuclease treatment. This implied that at least some of the inverted repeats were separated by a spacer sequence not part of the inverted repeat. The spacer would form an S1 sensitive loop at the end of the hairpin. S1 would also remove sequences outside the inverted repeat (Fig. 4.11).

The experiment was repeated with DNA sheared to a length of 0.3 kb and then only 6% of the DNA bound to hydroxyapatite after zero time reassociation. The reduction was due to loss of sequences with a larger spacer between the repeats, to loss of the ends of very long inverted repeat sequences and to loss of sequences outside the inverted repeat. With the short DNA, 3% of the original DNA also survived the S1 nuclease treatment, so the loops and ends associated with each inverted repeat were shorter. The precision of the base pairing in the S1 nuclease treated inverted repeats was checked by thermal denaturation and the T_m, after correction for DNA length, was $1.6°$ below that of native DNA, which implies almost no mismatching.

The lengths of the inverted repeats isolated as above, including S1 nuclease treatment, were measured by electrophoresis in polyacrylamide gels or by ultracentrifugation and found to range from about 0.1 kb to about 1 kb with an average of about 0.2 kb. These lengths are much shorter than the original DNA length of 18 kb so they probably give a good picture of the *in vivo* distribution. Very short sequences (<50 b) would not be detected because they would not bind to hydroxyapatite. The figures imply the existence of about 2 million inverted repeat sequences in the human haploid genome. S1 nuclease treatment of inverted repeats removes the loops of single stranded DNA connecting the two inverted copies of the repeat so they will no longer reassociate at 'zero time' if the reassociation experiment is repeated. The reassociation experiment was carried out on the purified inverted repeats and 10% reassociated at zero time. This represents sequences where the inverted repeats are adjacent, so that no loop is formed, and also sequences which contain an internal inverted repeat. The rest of the inverted repeats reassociated at a range of $c_0 t$ values up to about 10, showing that most inverted repeat sequences are repeated at least 1000 times in one cell and in many cases much more than 1000 times.

In situ hybridization to metaphase chromosomes showed that the inverted repeat sequences were distributed at many sites.

The function of inverted repeats is unknown. An interesting observation has been made in the study of *Drosophila* rDNA (see section 4.11). A non-coding sequence has been found to be inserted into the 28S rRNA coding sequence of some of these rDNA repeats from the X chromosome but not the Y chromosome, and many of these inserted sequences contain an inverted repeat sequence at the ends, as if the inverted repeat were involved in the translocation of DNA

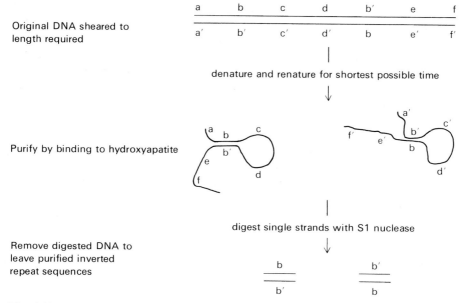

Fig. 4.11 Isolation of inverted repeat sequences.

sequences from one part of the chromosome to another part of the same chromosome.

Inverted repeat sequences have also been observed at specific protein binding sites in prokaryotes, as in Table 4.1.

An inverted repeat sequence may be termed a palindrome since it appears identical when viewed from either end, as can be seen from the table and the

Table 4.1

Lac Operator

   ...TGTGTGG̅AATTGTG̅AG̅CGGA̅TA̅ACAATTT̅CACACA...
   ...ACACAC̲C̲TTAACAC̲TCGCC̲TA̲TTGTTAAA̲GTGTGT...

CAP Site

         ...TGAGT̅T̅AG*CTC̅ACTCA...
         ...ACTCAATC*GTGAGAGT...

λ Operator O_L1

         ...TA̅T̅CACC̅GC̅C̅C̅A̅GTGG̅TA...
         ...A̲TAGTGGCGGTCAC̲C̲AT...

Sequences noted above were obtained from Gilbert and Maxam (*LAC*), Maizels (*CAP*) and Ptashne (*λ*). * = center of rotational symmetry; __ = non-identical base pair.
   (From Leder P., Honjo T. & Swan D. (1977). *The organization and expression of the eukaryotic genome* (eds) Bradbury E. M. & Javaherian K. Academic Press, London.)

fact that the two strands have opposite 3'–5' directions. The other property of an inverted repeat sequence is that half of one strand is complementary to the other half of the same strand so that when the strands are separated each half 'snaps back' to form a double stranded structure with itself (hence the 'zero time reassociation' property). It has been pointed out by Gierer that the strands could do this within the chromatin, given some other stabilizing condition such as specific protein binding. It is not known if such structures exist. In the case of the short inverted repeats recognized by restriction nucleases, it seems to be the symmetry of the dimer enzyme structure which leads to the recognition of a symmetrical DNA sequence.

4.5.2 Simple sequence DNA

Simple sequence DNA contains a short nucleotide sequence, up to about 10 nucleotides, that is repeated many thousands of times to produce a long DNA molecule. Such DNA molecules occur in many, if not all, eukaryote cells. There may be either one major type of simple sequence DNA, as in mouse for example, where about 10% of the DNA contains consecutive repetition of a pentanucleotide sequence or there may be several simple sequence DNAs, as in guinea pig for example. The very high degree of repetition, up to 1 million copies per cell as in mouse, means that in DNA reassociation experiments the simple sequence DNA reassociates very rapidly, with $c_0 t$ values in the range 10^{-3}–10^{-1} mol s 1^{-1}. Consequently simple sequence DNA may be known as rapidly reassociating DNA. This property also provides a method of preparing simple sequence DNA, although it is not suitable for large scale preparation since the reassociation must be carried out at low concentration in order to achieve sufficiently low $c_0 t$ values.

Simple sequence DNA behaves anomalously in other respects, such as spectroscopic properties and density in solution. This behaviour is due to the fact that the nearest neighbour frequencies for nucleotides in simple sequence DNA are usually very different from those in a complex DNA sequence. Consequently, the density in solution and the thermal denaturation temperature of a simple sequence DNA may be different from the equivalent properties of a complex DNA with the same base composition. In addition, of course, the simple sequence DNA may have a base composition different from that of the rest of the DNA in the cell. These differences form the basis of the most widely used method for the preparation of simple sequence DNA—density gradient centrifugation. In the case of mouse, for example, the bulk of the DNA forms a fairly broad band in a CsCl density gradient but the simple sequence DNA forms a small sharp band on the high density edge of the broad band. Such bands are called 'satellite' bands and the term 'satellite DNA' has sometimes been used to describe simple sequence DNA. However, that term should be avoided because other types of DNA can also form satellites in CsCl gradients e.g. ribosomal genes, histone genes or mitochondrial DNA.

The function of simple sequence DNA is unknown. In some cells it has been shown, by *in situ* hybridization, that simple sequence DNA is concentrated in the centromeres of chromosomes. Although this is a well established finding, simple sequence DNA may occur elsewhere in the chromosome. Simple sequence DNA is sometimes referred to as 'centromeric DNA' but of course this term should be restricted to that class of simple sequence DNAs that occurs in centromeres.

In situ hybridization is a general technique for studying the position of specific DNA sequences. First, the DNA sequence in question is purified in a radioactive form. Then the cells or chromosomes are prepared and fixed for microscopy, the fixed DNA being denatured *in situ* (for example with 0.1 N NaOH) and allowed to renature in the presence of the specific radioactive sequence, which then hybridizes wherever its complementary sequence occurs. After renaturation, the excess radioactive DNA is washed away and the sample prepared for autoradiography. After autoradiography the sample is examined, usually by light microscopy, and the positions of radioactivity as revealed by spots on the autoradiograph indicate the position of the sequence being studied, for example Fig. 4.12.

It is also known that the amount of simple sequence DNA, as a proportion of the total, is reduced in cells containing giant polytene chromosomes compared with normal diploid cells. Simple sequence DNA is not transcribed into RNA so we can guess that such DNA has a structural role, as yet unknown, in the structure or organization of eukaryote chromosomes.

EVOLUTION IN A SIMPLE SEQUENCE SATELLITE DNA

Satellite DNA was a natural first target for DNA sequencing studies and a substantial amount of direct sequence information is now available. This information has provided rather little help in determining the function and mode of operation of simple sequence DNA, unlike the situation in the case of sequences in and around genes. However, the satellite sequences have been used to form the basis of proposals for evolution of DNA sequences. For example, consider the sequence of bovine 1.706 satellite DNA. It has a long range repeat of 2350 bp which is divided into four segments. Each segment consists of different variants of a basic 23 bp sequence which is itself composed of a 12 bp sequence and a related 11 bp sequence. The complete satellite appears to have evolved from the 11 bp or 12 bp sequence mostly by duplication with some deletions which seem much more common than point mutations or insertions. The following, more detailed, scheme has been proposed. The basic unit is the 12 bp sequence. This was duplicated to give 24 bp and then one bp was deleted to give the 23 bp repeat which occurs throughout the satellite. The occasional variations in the 23 bp repeat are not randomly distributed within the satellite, in this case, so it has been proposed that the sequence 'grew' by additional duplications to 46 bp, then 92 bp and then a complete duplication of the 92 bp

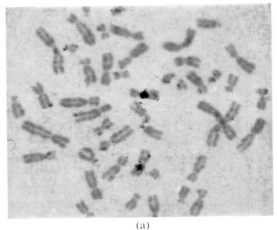

(a)

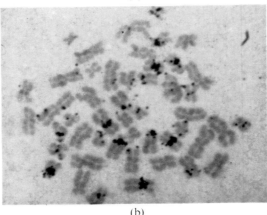

(b)

Fig. 4.12 *In situ* hybridization. (a) and (b) are micrographs of human (a) or chimpanzee (b) chromosomes. The chromosomes have been hybridized to radioactive human satellite III DNA. The black dots on the pictures show the regions where hybridization has occurred. The hybridization in (a) occurs at the centromere of chromosome 9. The occurrence of hybridization in (b) shows that the sequence of human satellite III has been conserved during evolution from chimpanzee to human. (From Jones K. W. *et al.* (1973) *Chromosoma Berl.* **42**. 445.)

plus reduplication of one of these 92 bp sequences to give a 276 bp (3×92) repeat. The next step was 'expansion' of the end 23 bp of the 276 bp repeat leaving the original 253 bp (276–23) plus a large number, n, of repeats of the 23 bp sequence. Then 2 bp were deleted from the 253 bp unit and the whole sequence of ($253 + 23 n$) bp was duplicated. This was modified by expansion or contraction of the array of n 23 bp units, deletion of 4 bp from one of the 253 bp sequences and generation of some 22 bp units. Finally, the 2530 bp sequence was completed and at that time it was duplicated 50 000 times to give the satellite DNA now observed.

Somewhat different sequence arrangements, more like a homogeneous randomly diverged sequence, are observed in many other satellites. The mechanism of evolution is unknown but there is no compelling evidence for selection in the evolution of satellite sequences. The sequences could have been developed by random processes such as unequal crossing over, (see Chapter 5).

The particular example described above shows similarities when compared with the sequences found in the repeating units of *Xenopus* 5s rRNA genes. The conservation of the structural gene sequence in this case shows evidence of selective pressure during evolution but the spacer sequences appear to have evolved randomly.

4.5.3 Middle repetitive DNA

This is a heterogeneous class of DNAs consisting of sequences which are repeated between 10 and 1000 times per genome. Middle repetitive DNA can be isolated by DNA reassociation experiments as DNA which reassociates at intermediate c_0t values, approximately $1-100$ mol l^{-1}. A number of functions for middle repetitive DNA have been proposed and there can be no doubt that different DNAs within the class will have different functions. The sequences are relatively long, 1 kb–30 kb, before repetition and are not necessarily adjacent—they may be widely spaced on the chromosome, unlike simple sequence DNA where the repeats are clustered.

The most clearly documented function is that of the repetitive gene. The genes for the large ribosomal RNAs, 5S ribosomal RNA, transfer RNA and histones all fall within the class of middle repetitive DNA.

It seems likely that another major function of middle repetitive DNA concerns the control sequences in DNA. It is postulated that specific DNA sequences can be recognized by molecules, probably proteins, that initiate or prevent the initiation of RNA synthesis at particular sites. These control sequences would be repetitive because one initiator or repressor would be expected to affect a number of different genes. It is interesting that genes occuring together in an operon in bacteria may be very widely spaced in eukaryotes, implying the necessity of a repetitive control sequence to provide co-ordinated control of those genes. Moreover, as discussed in Chapter 3, some widely spaced genes in bacteria, making up a regulon, have common control sequences which are sensitive to the same regulator. (It should also be remembered, however, that the histone genes may provide an example of an operon in those eukaryotes where the genes are coded on the same strand, as in sea urchins.) Other types of control sequence are possible. In many cases the RNA molecule that is initially transcribed from the DNA may contain a much longer sequence than that which occurs as mRNA in the cytoplasm. The enzymes that break, digest and rejoin RNA molecules during their maturation in the nucleus may recognize specific sequences or secondary structures that occur in the initial RNA product of many genes. Other repetitive sequences may be involved in determining the stability of mRNA molecules. These possibilities are supported by the fact that middle repetitive sequences are usually interspersed with genes.

Another, very speculative proposal is that much of the 'excess' DNA in the cell is a pool of unstable genes that can mutate rapidly in a 'search' for better genes which will be selected for when they occur by chance. And of course it is possible that the 'excess' DNA actually is redundant and represents sequences no longer used. On this basis small viruses with overlapping genes may represent recently evolved systems that have become much more efficient in their use of DNA.

4.5.4   Single Copy DNA

About half the human genome is made up of a large group of nucleotide sequences containing no significant repeats called single copy DNA sequences which are interspersed with middle repetitive DNA sequences. Some single copy DNA sequences represent single copy gene sequences but since the number of genes in a mammalian genome is only 10^4–10^5 the amount of DNA in single copy gene sequences is, at most, $10^5 \times$ length of an average gene, i.e. about 10^8 bp all together. This is only a small fraction of the total amount of single copy DNA per genome. The function of the other single copy DNA sequences is unknown although much of it is probably associated with single copy genes and some is even transcribed into RNA. However these sequences are not found in messenger RNA and hence are not translated into protein.

The study of single copy sequences was initially limited to those genes that produce an abundant mRNA so permitting its purification and use as a probe for the DNA sequence. The DNA sequence could then be cloned and studied in detail, as in the case of the gene for ovalbumin. However, with the more widespread use of cloning techniques, information on other types of single copy DNA sequence is becoming available. The potential interest of these studies can be seen in the work on the ovalbumin gene (section 4.14).

4.6   DETERMINATION OF THE NUCLEOTIDE SEQUENCES OF NUCLEIC ACIDS

The first complete nucleotide sequence of a naturally occurring nucleic acid, that of alanine transfer RNA, was published in 1965. At that time there were major difficulties in purifying the nucleic acid and in piecing together the many fragments obtained by nuclease digestion of the nucleic acid. These problems have been enormously reduced by the use of modern techniques particularly DNA cloning, restriction enzymes and gel electrophoresis. Cloning of DNA sequences provides an excellent method of purification.

Restriction enzymes are used to generate a manageable number of fragments and the variety of restriction enzymes available together with the characterization of the DNA fragments by gel electrophoresis makes it easy to arrange the fragments in the correct order. Finally, stretches of sequence up to 500 nucleotides long can be determined directly.

4.6.1   Restriction Enzymes

This is a group of enzymes found in bacteria. Each type of restriction enzyme is one of a pair of enzymes; the other member of a pair is a sequence specific methylating enzyme. After the cellular DNA is replicated *in vivo* the methylating enzyme recognizes a specific sequence of nucleotides and modifies the DNA by methylating one of the nucleotides of the sequence. For example, some strains of *E. coli* contain a methylating enzyme that recognizes the sequence . . . GA ATTC . . . on the newly synthesized DNA strand and methylates the starred adenosine. Since this sequence is identical to its complementary copy the methylating enzyme recognizes and methylates the same position on the other new daughter strand. Methylation protects the sequence from the restriction enzyme that recognizes the same sequence. If the sequence is not methylated, as on an invading viral DNA, then the restriction enzyme cuts the unmodified DNA either between two specific nucleotides in the sequence (type II enzymes) or at an unspecified point (type I enzymes). In our example from *E. coli* the restriction enzyme is known as *Eco* RI and cuts at the point indicated by the arrow: . . . G↓AATTC . . . . A pair of enzymes constitutes a restriction-modification system. Such a system has the effect of protecting the host DNA, which is modified, and attacking foreign DNA which is unmodified, hence restricting the host-range of the invading DNA. Restriction enzymes have become a remarkably important tool for the study of DNA sequences i.e. for molecular genetics.

Consider a random sequence of nucleotides containing 20% G, 20% C, 30% A and 30% T i.e. equivalent to a DNA of 40% G + C like mammalian DNA. The probability of G occurring at a given site is 0.2; the probability of A occurring next is 0.3 and so on. The probability of GAATTC occurring is thus $0.2 \times 0.3 \times 0.3 \times 0.3 \times 0.3 \times 0.2$ which is 0.000324. Hence, on average, this sequence will occur once every 3086 nucleotides. Thus the restriction enzyme *Eco* RI would be expected to cut mammalian DNA into specific fragments of average size approximately 3 kb. Any particular fragment, of course, may be much smaller or much larger and mammalian DNA sequences are not random, but this gives an idea of the size obtained. Larger fragments can be obtained by partial digestion and smaller fragments by digestion with other restriction enzymes. Many restriction enzymes recognize a tetranucleotide sequence and these produce smaller fragments with a single digestion. In most cases restriction enzymes can be used to generate fragments suitable for direct sequence analysis. Larger sequences can be built up from adjacent or overlapping fragments. The fragments can usually be arranged in their correct order before direct sequencing is undertaken.

Initial analysis of a restriction nuclease digest of a particular DNA is usually carried out by separating the DNA fragments according to their length by electrophoresis. At the pH value used (about pH 8) the charge on the DNA fragment is independent of its base composition and proportional to its length i.e. the number of phosphate groups. In gels the migration of the DNA is also governed by the steric hindrance of the gel structure which reduces the DNA mobility as the DNA length increases. This dominates the charge effect under the conditions normally used so that the electrophoretic mobility of DNA molecules in gels decreases as the length increases.

In a gel electrophoresis experiment a slab of agarose and/or acrylamide gel is cast with pockets for samples. The samples are electrophoresed in parallel lanes or tracks which allow for easy comparison between tracks. For short DNA molecules (hundreds of base pairs) each possible length can be separated. For longer DNA molecules this is not normally achieved but DNA lengths can still be determined with considerable accuracy if the gel is calibrated using a set of DNA fragments of known length in one of the tracks of the gel. The known lengths are obtained by sequencing.

For preparative purposes the gel can be cut into slices and DNA fragments eluted from the slices.

4.6.2   Direct Sequencing

One important method of direct sequencing of DNA uses DNA polymerase with high resolution electrophoresis in polyacrylamide gels and a sequence of several hundred nucleotides can be read from one such gel. It is necessary to start with one strand of the DNA fragment to be sequenced, together with a small fragment complementary to one end. These fragments can usually be prepared using restriction enzymes, gel electrophoresis and, maybe, density gradient centrifugation. Figure 4.13 shows the starting material. The small fragment is reassociated with its complementary sequence at one end of the fragment to be sequenced so providing a short stretch of double stranded DNA.

This provides a substrate for DNA polymerase I which 'repairs' the DNA by synthesizing a new strand of DNA on the single strand provided. Synthesis begins at the double stranded region which acts as a primer and continues along the single strand which is used as a template for synthesis of a new complementary strand. DNA polymerase synthesizes the complementary strand by adding a nucleoside 5' phosphate to the free 3'-OH of the primer leaving a new 3'-OH ready for the next nucleoside 5' phosphate as indicated schematically in the example in Fig. 4.14.

The new DNA is radioactively labelled by using ^{32}P labelled nucleoside triphosphates with the label in the α-phosphate position. Under appropriate conditions this reaction will continue for many hundreds of nucleotides along the template. For sequencing purposes it is required to stop the polymerization at specific nucleotides. This is achieved by including a dideoxynucleoside triphosphate, i.e. a deoxynucleoside triphosphate lacking the 3'-OH group (Fig. 4.15). In our previous example, if the next nucleotide inserted were dideoxy-guanosine-phosphate then the polymerization would stop (Fig. 4.16), since there would be no 3'-OH group to attach the next nucleotide to. For sequencing purposes four DNA polymerase reactions are carried out with the same DNA molecules but a different dideoxynucleotide in each reaction. Then all the newly synthesized molecules from the reaction with dideoxyguanosine triphosphate (ddGTP) will terminate with dideoxyguanosine (ddG); all the DNA molecules from the reaction mixture containing ddATP will terminate in ddA and so on. The conditions are arranged so that ddGTP, for example, is only used occasionally so that the newly synthesized DNA molecules will be of all possible lengths ending in ddG. Similarly, the ddATP reaction will have all possible DNA lengths ending in ddA and so on.

The complex mixture of newly synthesized DNA molecules from each reaction is loaded onto an acrylamide gel and separated by electrophoresis. The resolution is sufficient to separate molecules differing in length by one nucleotide. The newly synthesized molecules are radioactive and are detected by autoradiography of the gel. Every possible length of DNA (up to a maximum of several hundred nucleotides) will be represented on the autoradiograph. Those ending in ddG will appear on the gel track from the reaction with ddGTP; those ending in ddA will come from the reaction with ddATP and so on. Consequently, the sequence complementary to the original template DNA can be simply read from the autoradiograph (Fig. 4.13).

Other methods have been developed to generate labelled fragments ending in a particular nucleotide and these are also used for sequencing studies. In particular, the methods of Maxam and Gilbert use DNA labelled at one end. This can be achieved, for example, by removing the terminal 5'-phosphate using alkaline phosphatase and replacing it with ^{32}P labelled phosphate using polynucleotide kinase.

Only one end needs to be labelled so the other end is removed, for example using a restriction enzyme to leave a double stranded DNA labelled at one end. This DNA is cleaved in four separate partial reactions by chemical methods with specificity for particular nucleotides. These reactions generate DNAs of all possible lengths which are separated by gel electrophoresis and those including the end of the DNA are detected by autoradiography. The sequence can then be read from the autoradiograph as in the method described above. Similar principles are used for sequencing RNA molecules directly but RNA sequences may also be determined indirectly by converting the sequence to a DNA sequence as in the sample given in the next section and Fig. 4.20.

4.7 GENETIC ENGINEERING *IN VITRO* AND MOLECULAR CLONING

A number of enzymes has become widely used in the last 10 years for manipulating nucleic acids in solution. Among the most important are: RNA dependent DNA polymerase; DNA polymerase I; restriction enzymes; DNA ligase. There are other useful enzymes such as polynucleotide kinase which is used for labelling the 5' end of DNA molecules; exonucleases which remove nucleotides from the ends of nucleic acids; S1 nuclease which digests single

(a) Synthetic method (enzymatic)

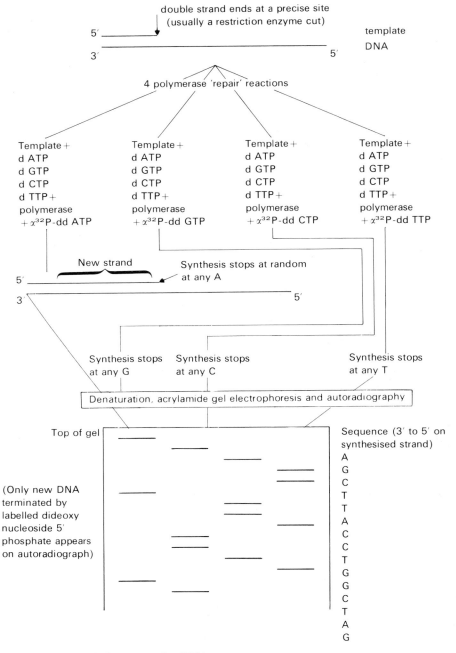

Fig. 4.13   Methods for sequencing DNA

(b) Degradation method (chemical)

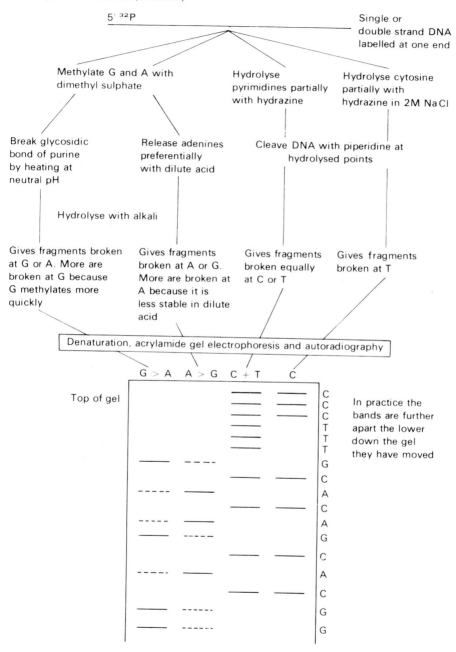

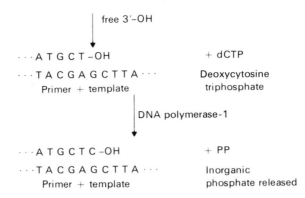

Fig. 4.14   Action of DNA polymerase-1.

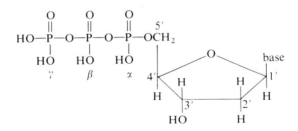

2'-deoxynucleoside triphosphate (normal precursor)

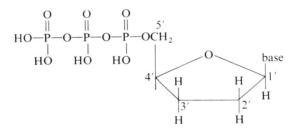

2',3'-dideoxynucleoside triphosphate (chain terminating precursor)

Fig. 4.15   The structure of the normal and chain terminating nucleoside triphosphates used by DNA polymerase-1.

stranded DNA but not double stranded DNA; and non-specific endonucleases which have been widely used to analyse chromatin structure.

4.7.1   RNA dependent DNA polymerase (also known as reverse transcriptase) This enzyme is used *in vivo* by viruses whose genetic information is carried in the virus particle as RNA. The RNA genome is converted to DNA ready for

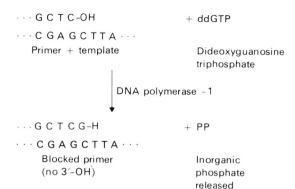

Fig. 4.16 Chain termination by a dideoxynucleoside.

insertion into the host genome. *In vitro* it is used to synthesize a DNA strand complementary to an RNA strand, such as a messenger RNA. This new strand is called a cDNA. The DNA can then be cloned.

4.7.2. DNA Polymerase I

This enzyme is used *in vivo* to repair damaged DNA. It can both remove nucleotides from one strand of a DNA molecule and replace them. The double action is used to label DNA with ^{32}P to very high specific activity by a process called '*nick translation*'. Single strand breaks (called nicks) are introduced into the DNA chemically to give points of damage where DNA polymerase I can start. It then moves along the DNA removing nucleotides and replacing them as it goes (called nick translation) (Fig. 4.17). The new nucleotides can be

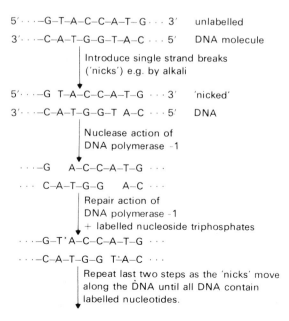

Fig. 4.17 Labelling DNA *in vitro* by 'nick translation'.

radio-actively labelled. DNA polymerase I has another important role, namely to synthesize the complementary strand to the cDNA prepared by RNA dependent DNA polymerase to give a double stranded cDNA.

4.7.3 Restriction Enzymes

These enzymes are used for sequence studies (q.v.). The recognition sequences for restriction enzymes are palindromes and in many cases the DNA is cut to leave short single stranded ends. These ends will reassociate under appropriate conditions either with their original partner or with any other DNA that has also been cut by the same restriction enzyme. These ends are called 'sticky ends' or 'cohesive ends' or 'overlapping ends'. For *Eco* RI the overlapping ends are (Fig. 4.18).

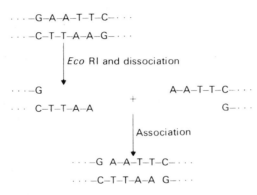

Fig. 4.18 Breakage and rejoining of DNA molecules using the 'sticky ends' produced by the restriction enzyme *Eco* RI.

4.7.4 DNA Ligase

This enzyme is used *in vivo* for DNA replication. It joins DNA molecules *in vitro* which have reassociated with overlapping ends (Fig. 4.19).

These, and other, enzymes are used to construct novel DNA molecules by splitting and recombining them in new ways. This has led to the technique of molecular cloning. A DNA molecule constructed by splitting, rearranging and re-joining is known as a recombinant DNA molecule. Where the original DNA molecules are complex the mixture of recombinant DNAs is very heterogeneous. For molecular cloning recombinant DNA is made from the DNA to be cloned and a DNA which will replicate in a suitable host. The latter DNA is known as

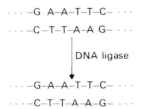

Fig. 4.19 Sealing DNA molecules together with DNA ligase.

the *vector* and is usually a plasmid or a virus such as modified λ or SV40. Suitable vectors will replicate in their host even when they are recombined with another DNA sequence.

Recombinant DNA molecules, then, are introduced into the host appropriate to the vector part of the recombinant DNA. This can be done directly, by the process of transfection, or more efficiently in the case of a λ phage vector by packaging the DNA into the phage coat *in vitro* and then infecting the host, *E. coli*. Recombinant DNA molecules that contain all the necessary parts of the vector will replicate in the host. This provides almost unlimited amplification of the recombinant DNA sequence to produce a very large number of copies which can be studied biochemically. The original DNA can be recovered from the vector by cutting the recombinant DNA with the appropriate restriction enzyme.

In the case where there are many different viable recombinant DNA molecules they can be separated by spreading the transformed or infected cells on a nutrient agar plate so that each individual recombinant DNA replicates independently of the others and forms a separate colony or clone on the plate. Individual clones can then be picked and grown up in bulk. Various selection and assay methods are available for choosing the colonies of interest. This process, called cloning, allows an individual DNA molecule to be isolated from a heterogeneous mixture and then grown up in large amounts.

These procedures have additional potential in that it is possible for the DNA sequence inserted into the vector to be transcribed into RNA and translated into protein. The DNA sequence will be translated in a prokaryote host if it contains a promoter for transcription and an initiation site for translation or if these sequences are provided by the vector. In this way proteins present in mammalian cells can, in principle, be synthesized in bacterial cells by cloning the mRNA sequence from the mammalian cells in a vector that grows in bacteria. There are practical difficulties if post-translational processing is required, for example, and in obtaining correct initiation and termination, but expression of eukaryotic sequences in bacteria has been observed using these techniques. This suggests the possibility of large scale production of mammalian proteins or polypeptide hormones of medical importance.

As an example of the application of these techniques consider the cloning of the mRNA sequence coding for insulin in rats (Refer to Fig. 4.20). In the first stage the mRNA was incubated with RNA dependent DNA polymerase and deoxynucleoside triphosphates to give a double stranded RNA–DNA hybrid. Under these conditions a small loop of DNA appeared at one end. In stage 2 the RNA part of the hybrid was removed by digestion with alkali. RNA dependent DNA polymerase was then used again but this time it copied the single stranded DNA to give a complete DNA double strand. The loop of DNA at one end was removed by digestion with S1 nuclease, stage 4, to give insulin cDNA. In stage 5 small synthetic oligodeoxynucleotides carrying the recognition

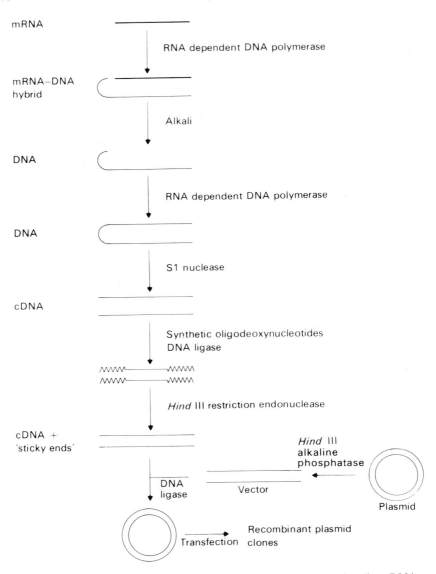

mRNA

RNA dependent DNA polymerase

mRNA–DNA hybrid

Alkali

DNA

RNA dependent DNA polymerase

DNA

S1 nuclease

cDNA

Synthetic oligodeoxynucleotides
DNA ligase

Hind III restriction endonuclease

cDNA + 'sticky ends'

Hind III alkaline phosphatase

DNA ligase

Vector

Plasmid

Transfection

Recombinant plasmid clones

Fig. 4.20 Schematic representation of the main stages in cloning an insulin mRNA sequence in a plasmid.

site for the restriction enzyme *Hind* III were added and the mixture incubated with DNA ligase. This joined the oligodeoxynucleotides to the ends of the insulin cDNA. The composite DNA was then incubated with *Hind* III which does not cut insulin cDNA because there are no recognition sites. This reaction, stage 6, produced cDNA with 'sticky ends' from the *Hind* III digestion of the synthetic oligodeoxynucleotide. Next, the vector was added. This was a plasmid, pmb9, which contained only one recognition site for *Hind* III. The plasmid had been incubated with *Hind* III which cut the circular DNA once to give a linear molecule and with alkaline phosphatase which removed the 5′ phosphate groups from the ends of the molecules. This removal prevents the action of DNA ligase. The vector was allowed to recombine with the cDNA through the 'sticky ends' left by *Hind* III and the recombinants were sealed with DNA ligase. (Only recombinants containing cDNA could be sealed because they contained terminal 5′ phosphates.) The recombinant plasmids were cloned by transfection into *E. coli*.

4.8 HISTONE GENES

The genes for histones are repeated within one genome, many hundred fold in sea urchin for example. Histone mRNAs can be obtained relatively easily from cleaving sea urchin where histone synthesis is a major cellular activity. This allows the histone coding DNA to be recognized as a satellite in equilibrium CsCl density gradients. Figure 4.21 shows that the histone coding DNA forms a band within the main band of total DNA. However, it could be separated by the use of the drug actinomycin D that binds specifically to G–C-rich DNA and reduces its density in CsCl.

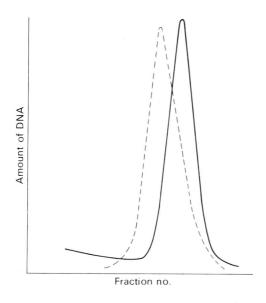

Fig. 4.21 Detection of genes for histones as a cryptic satellite in sea urchin DNA.

- - - - - Hybridisation with histone mRNA
————— Total sea urchin DNA

The isolated DNA was analysed using restriction enzymes that showed there was one fragment of 5.58 kb which hybridized to all five histone messengers. The fact that histone coding DNA formed a distinct satellite in CsCl gradients of high molecular weight DNA showed that these 5.58 kb repeats were clustered in one or a few large blocks. In *Drosophila*, *in situ* hybridization of histone mRNAs to polytene chromosomes confirmed that the genes were linked and clustered to one region of the chromosomes.

The entire gene cluster from sea urchin (Fig. 4.22) has been cloned and analysed by a variety of methods including direct sequencing.

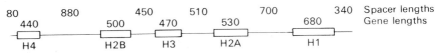

Fig. 4.22 Histone gene cluster of sea urchin *Psammechinus miliaris*. Each box shows coding DNA with its length, in base pairs, above and the histone for which it codes below. Each solid line shows non-coding DNA with its length, in base pairs, above. The total repeat is 5.58 kb.

Sequence analysis has shown that there are no inserted sequences of non-coding DNA within each individual histone gene. In sea urchin the five different genes occur with the same polarity along the DNA so they could be transcribed as a single unit. However, in *Drosophila* this is not the case and some genes are coded on one DNA strand and some on another.

4.9 ADENOVIRUS DNA

Adenovirus, discussed in more detail in Chapter 3, is a small virus that can be grown in human cell cultures and its DNA has been used as a probe for the study of the mechanisms of transcription and replication of the host DNA in human cell cultures. It is known, for example, that adenovirus DNA is transcribed by the host RNA polymerase II (B) and it seems likely that other features of the transcription process are common to adenovirus DNA and the normal host DNA. As with other viruses, adenovirus has the advantage that its specific mRNAs are present in high concentration in infected cells and these mRNAs can be hybridized to pure adenovirus DNA.

Adenovirus DNA is cut by restriction enzymes and *Eco* RI, for example, cuts adenovirus DNA into six fragments. The positions of the sites of attack by *Eco* RI and other restriction enzymes have been determined and Fig. 4.23 shows the 'restriction enzyme map' of adenovirus with the sites of *Eco* RI attack.

The fragments produced by *Eco* RI are of different sizes, a–f in Fig. 4.23, and

Fig. 4.23 Adenovirus DNA showing the sites of attack by restriction nuclease *Eco* RI.

they can be separated by agarose gel electrophoresis. Each mRNA produced by transcription of adenovirus DNA during growth can be isolated and hybridized to the separate fragments of adenovirus DNA. The results will show which parts of adenovirus DNA are transcribed and where each mRNA comes from in the adenovirus DNA sequence. Such an experiment was carried out with a specific adenovirus mRNA, the mRNA coding for a specific polypeptide called hexon. This hexon mRNA hybridized to *Eco* RI fragments a and b but not to the other fragments showing that the hexon gene lies in the region of adenovirus DNA defined by *Eco* RI fragments a and b.

The hybrid molecule formed from the *Eco* RI fragment a of adenovirus DNA and hexon mRNA was then studied in more detail by electron microscopy. The DNA and RNA were mixed and incubated under conditions where DNA–DNA double strands fall apart but DNA–RNA double strands are stable. The hybrid molecules formed during incubation were spread on a protein monolayer and examined by electron microscopy. In the electron micrograph single strands can be seen as thin lines and double strands as thick lines. The original expectation was that the DNA would be single stranded at its 5′ end, become double stranded (hybridized with hexon mRNA) near the 3′ end and have a 'tail' of hexon mRNA at the 3′ end as in the following scheme, (Fig. 4.24).

This technique is sometimes called *heteroduplex mapping*. The expected features were seen on the electron micrograph but there was an additional unexpected feature. This was hybridization of hexon mRNA to DNA sequences near the middle of the *Eco* RI fragment a and looping out of three regions of the DNA that did not hybridize. One such molecule is shown in Fig. 4.25.

The 5′ end of the DNA is single stranded and begins at the top of Fig. 4.25. If we trace along the DNA we come to a short double stranded region; then a

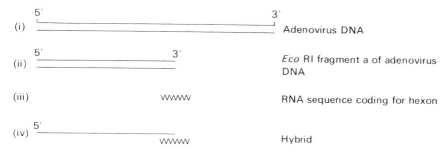

Fig. 4.24 Predicted hybridization of mRNA coding for hexon with fragment a of adenovirus DNA.

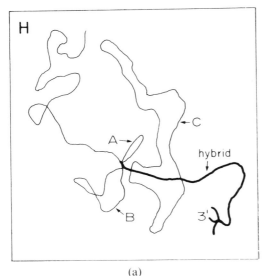

(a)

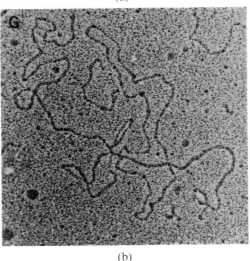

(b)

Fig. 4.25 Electron micrograph of a hybrid of hexon mRNA and a single stranded fragment of an adenovirus DNA. In the tracing, (a), the hybrid region is indicated by a thick line; loops A, B and C are sequences of single stranded DNA joined by RNA–DNA duplex regions resulting from the hybridization of DNA sequences from three regions on the left end of the genome to the 5′ tail sequences. The collapsed structure at the 3′ end of the hybrid is pure RNA that overlaps the end of this DNA fragment. (From Berger S. M., Berk A. J., Harrison T. & Sharp P. A., *Cold Spring Harbor Symposium on Quantitative Biology* **XLII**, 525.)

loop, labelled A, of single strand DNA; then another short double stranded region; then another loop, B, of single strand DNA; then a third double stranded region; then a third loop, C, of single strand DNA; then a long double stranded region; and finally a complex 'trail' which is folded mRNA. The double strand regions are DNA–mRNA heteroduplexes and they show that the mRNA is complementary to sequences on the DNA. The single strand loops are DNA sequences not found in the mRNA and they show that the mRNA sequences come from four different sites on the DNA. The experiment is shown diagrammatically in Fig. 4.26.

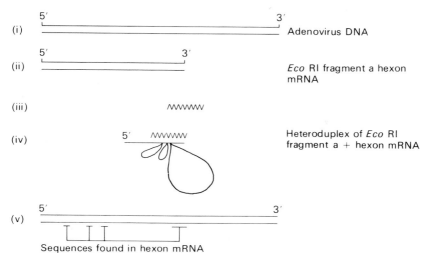

Fig. 4.26 Heteroduplex mapping of adenovirus hexon mRNA.

At the time of writing four mechanisms have been suggested to explain how the cell accommodates the lack of the continuous co-linearity of DNA and mRNA. (1) Each region of DNA could be transcribed separately and the RNA fragments joined by an RNA ligase enzyme. The difficulty here is in the improbability of the fragments coming together by chance so we have to postulate that the polymerase moves along the DNA continuously but transcribes discontinuously retaining the early transcript while moving on to the later section. There is however no evidence of this. (2) The transcription may take place on a modified DNA template where the intervening DNA has been deleted. There is other evidence that DNA sequences can be reorganized within a genome (see section 4.13) but not specific evidence in this case. (3) The DNA template could have a looped tertiary structure so that the sequences required in the mRNA were brought together ready for the polymerase. There is no evidence for this type of DNA structure nor for RNA polymerase being able to transfer from one DNA strand to an adjacent one. (4) A very large precursor RNA could be made by transcription proceeding from the first sequence required continuously through all the required sequences. The precursor RNA would then be 'processed' by splicing enzymes to give the final mRNA. This has the interesting feature that much of the control of mRNA formation is removed from the DNA to the area of RNA 'processing'. There is independent evidence from other systems that large precursor RNA molecules are made and a splicing enzyme has been isolated from yeast. Consequently, mechanism 4 is currently favoured.

4.10 GENES FOR 5S RIBOSOMAL RNA

The ribosomes of most eukaryotes contain four RNA molecules each, called 28S, 18S, 5.8S and 5S. The three larger rRNAs are derived from a single large precursor rRNA (45S in mammals) and the genes are clustered and located in the nucleolus. The genes for 5S RNA are also clustered but not closely linked to the genes for the larger rRNAs. Their arrangement has been studied in *Xenopus laevis*, which has many thousands of copies of the genes for 5S RNA in each cell. The DNA containing these genes (5S DNA) can be isolated as a satellite on CsCl gradients containing actinomycin. Digestion of this DNA with the restriction enzyme *Hind* III gives a single size of DNA fragment, 0.7 kb. Only a fraction of this, 0.18 kb, represents the sequence coding for 5S RNA and the remainder is 'spacer' DNA, partly transcribed into the 5S RNA precursor and partly non-transcribed. The bulk of the non-transcribed DNA is rich in dA + dT while the transcribed DNA is rich in dG + dC. This difference in overall base composition is a feature of many, but not all, gene arrangements. The spacer sequences have been exstensively studied and show substantial repetition rather like that in the bovine 1.706 satellite DNA (section on p. 135, Evolution in a Simple Sequence Satellite DNA). A map of the repeat is shown in Fig. 4.27 which also shows an intriguing feature termed a pseudogene. This is a second sequence almost identical with the first 100 nucleotides coding for 5S RNA that occurs as part of the dG + dC-rich sequence. Could this be a relic of evolution or does it have some other function?

4.11 GENES FOR LARGE RIBOSOMAL RNAs

As we have just stated eukaryote ribosomes contain 4 RNA species normally referred to as 28S, 18S, 5.8S, and 5S. In some lower eukaryotes 28S and 18S may vary slightly for example 26S and 19S in *Physarum* (Fig. 4.33). Studies on the genes for the large ribosomal RNAs were originally carried out on DNA from *Xenopus laevis* and from *Drosophila melanogaster*, by the techniques of hybridization and CsCl density gradient centrifugation. Saturation hybridization of separated 18S and 28S rRNA to total DNA was used to determine the amount of DNA complementary to these rRNAs in *Drosophila* mutants with 4, 3, 2 or 1

                                     5S RNA gene   Pseudo-gene

                    (A + T)-rich   (G + C)-rich

Fig. 4.27 Organization of the 5s rRNA gene repeat in *Xenopus*. (From Jacq C., Miller J. R. & Brownlee G. G. (1977) *Cell* **12**, 118.)

nucleolar organizers per cell and the amount of DNA complementary to rRNA was found to be strictly proportional to the number of nucleolar organizers. Similar results were obtained with *Xenopus* mutants containing 2, 1 or 0 nucleolar organizers and *in situ* hybridization, particularly in *Rhynchosciara* and *Sciara*, showed that rRNA complementary sequences are located in the nucleolar organizer regions of the chromosomes. Hybridization of rRNA to *Xenopus* DNA banded in a CsCl gradient, showed that the DNA complementary to rRNA banded as a satellite of higher density than the main DNA band. The amount of this satellite was also proportional to the number of nucleolar organizers and DNA isolated from nucleoli yields up to 90% satellite (in *Physarum*). The CsCl satellite exists even when the centrifuged DNA is of high molecular weight compared with the apparent length of the rDNA (i.e. the DNA coding for the rRNA precursor and its associated nontranscribed sequences). This shows that many rDNA repeats must be clustered together since the presence of main band DNA between the rDNA repeats would have dragged the rRDA satellite into the main band. At extremely high molecular weight the rDNA is integrated with the rest of the DNA (unlike the amplified rDNA in *Xenopus* oocytes and the extrachromosomal palindromes in *Physarum* and *Tetrahymena*). These conditions were dramatically confirmed by the electron micrographs of transcribing rDNA (Fig. 4.3) showing arrays of DNA sequences with nascent RNA chains attached, each separated by spacer DNA with no RNA attached. Transcription was in the same direction on adjacent rDNA repeats so the rDNA was repeated in tandem. Unfortunately such pictures cannot be used to determine the amount of DNA in the transcribed and spacer sequences since the packing ratios of the DNA in these sequences are unknown. Other electron microscope techniques, however, do allow such measurements to be made. They use partial denaturation or R-loops or secondary structure or hybrid DNA–RNA molecules.

Partial denaturation maps rely on the fact that regions of DNA of low dA + dT content appear to be able to 'melt' i.e. the strands separate, independently of the rest of the DNA. In many cases, the base compositions of coding and spacer regions of the DNA are sufficiently different so that an electron microscope sample can be prepared in which the G–C rich sequences remain double stranded and the A–T rich sequences form single stranded loops. The pattern of denaturation loops can be used to measure the DNA repeat length. A particularly fine example occurs with histone genes.

Still more information can be obtained from R-loop maps which rely on the additional fact that an RNA–DNA double strand is more stable than a DNA–DNA double strand of the same sequence under the same conditions. An R-loop is made by incubating double stranded DNA under conditions where it is partially denatured, as above, and adding RNA complementary to specific sequences of the DNA. The RNA forms a stable hybrid with any complementary DNA sequences that are denatured so that when the denaturing conditions are removed and the DNA–RNA mixture spread on an electron microscope grid the DNA appears as a double strand but with loops of single stranded DNA wherever an equivalent RNA sequence has displaced it (Fig. 4.28).

Secondary structure mapping is used with single stranded nucleic acids and relies on the characteristic pattern of the double stranded loop regions which occurs in many sequences, e.g. 28S rRNA. This pattern can be used to identify particular sequences on different molecules. For example, it has been used to confirm the structure and processing of the rRNA precursor in HeLa cells.

Hybrid DNA–RNA molecules (also called heteroduplex molecules) are prepared by incubating DNA and RNA under conditions where DNA–DNA double strands are unstable but DNA–RNA strands are stable. Then, in rDNA hybridized with rRNA, a hybrid molecule is formed with the coding strand of DNA being single stranded over the spacer sequences and DNA–RNA double stranded over the coding sequences. These regions can be readily distinguished if the hybrids are spread for microscopy by the 'ethidium bromide gene 32' technique. 'Gene 32' in this context is a protein produced by T4 infected *E. coli* cells and coded for by gene 32 on the phage T4 DNA. This protein binds specifically to single stranded DNA but not DNA–RNA double strands under the right conditions. The protein–DNA complex is then spread in the presence of ethidium bromide instead of the usual cytochrome. Hence, single stranded regions appear thick and double stranded regions are thin as in Fig. 4.30, (section 4.11).

These electron microscope techniques have confirmed the tandem arrangement of 18S, 28S rRNA sequences and spacer DNA in *Xenopus* and *Drosophila*. Restriction enzymes have been used more extensively to study the pattern of

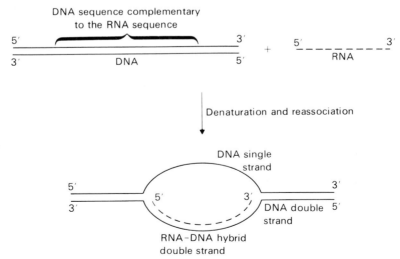

Fig. 4.28   Principle of R-loop mapping. (From White R. L. & Hogness D. S. (1977) *Cell* **10**, 177.)

rDNA repeats. Here the basic procedure is to digest either total DNA or purified satellite rDNA with one or more restriction enzymes and determine the molecular weights of the fragments that hybridize to rRNA. The results usually lead to a map of the rDNA showing the repeat length, the positions of restriction enzyme recognition sequences and the approximate positions of the rRNA complementary sequences as in Figs. 4.29 and 4.33.

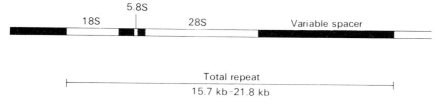

Fig. 4.29 Map of the repeated cluster of genes for ribosomal RNAs in *Xenopus laevis*. The open regions correspond to DNA sequences coding for the stable ribosomal RNAs, 18s, 5.8s and 28s. The shaded regions are spacer DNA sequences, some of which are not transcribed and some of which are transcribed but the RNA is removed during processing. The variable spacer region is composed mainly of tandem repeats of a 50 base pair sequence. The number of repeats varies, giving rise to variability in the length of this spacer.

Although the arrangement of rRNA genes in *Drosophila* shows overall similarity to that in *Xenopus*, there is one striking difference: some of the *Drosophila* DNA coding for 28S rRNA contains an insertion of variable length at a specific site in the 28S rRNA coding sequence. The function of this insertion is unknown. Figure 4.30 shows an electron micrograph of a hybrid molecule of rDNA and rRNA. The rRNA has hybridized to the sequences that code for it. Non-coding DNA appears as single stranded DNA (thick lines in this picture) and the coding DNA appears as a DNA–RNA hybrid (thin lines in this picture). The tracing, Fig. 4.30b, makes the picture easier to follow.

Data on the repeat unit of *Drosophila* rDNA is summarized in Fig. 4.1. As seen in Fig. 4.31 this structure is repeated in tandem. There is also a small 2S rRNA coding sequence near the 5.8S rRNA coding sequence.

The length of the non-transcribed spacer varies from about 2 kb to about 10 kb and the length of the sequence, I, inserted in the 28S coding sequence varies between 0.5 kb and 8 kb except in those repeats with no insertion. The small gap between 28S a and 28S b coding sequences is due to the elimination of 140 bases of the 28S rRNA after transcription. Genes containing the insertion are found on the X chromosome where they are apparently randomly mixed with genes containing no insertion. None of the ribosomal genes found on the Y chromosome contains the insertion.

Data on mammalian rDNA are less comprehensive than for *Xenopus* and *Drosophila*. However, studies on mouse, cow and human DNA are consistent with an rDNA arrangement in mammals of tandem repeats of alternating 18S

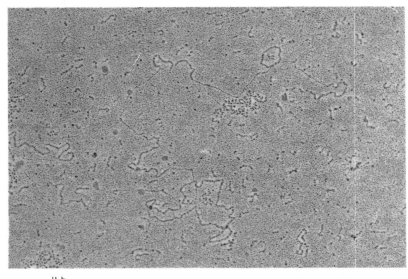

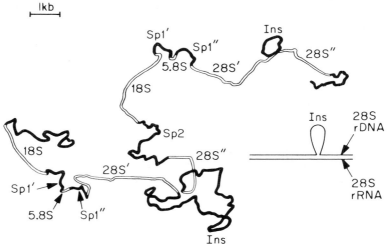

Fig. 4.30 Electron micrograph and tracing of an rDNA–rRNA hybrid molecule containing two gene sets each with an insertion in the 28s gene. Single strand DNA appears thicker than duplex DNA in these pictures. The inserted sequence in the 28s gene is marked Ins and the other spacer regions are marked Sp1, Sp2, Sp3. The small 'gap' in the 28s gene doesn't show up in this picture. (From Pellegrini M., Manning J. & Davidson N. (1977) *Cell* **10**, 216.)

and 28S coding sequences, with a transcribed spacer between the 18S and 28S genes, and a larger non-transcribed spacer between the 28S and 18S genes, as in *Xenopus* and *Drosophila*. There is heterogeneity of spacer lengths in mammals, as in *Xenopus* and *Drosophila*, but the average repeat length is greater in

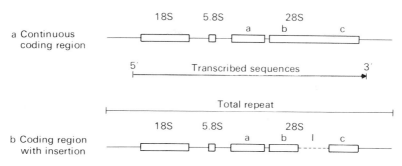

Fig. 4.31 The two different repeat units in Drosophila DNA coding for ribosomal RNA. (After White R. L. & Hogness D. S. (1977) *Cell* **10**, 177.)

mammals. For example the human rDNA repeat is at least 30 kb while the *Xenopus* rDNA repeat is 14 ± 3 kb. The data for mice and humans was obtained by digesting total DNA with restriction enzymes such as *Eco* RI, *Hind* III or *Bam* HI. The digested DNA was fractionated by acrylamide gel electrophoresis into bands of different molecular weight. The bands were denatured after electrophoresis and then transferred by electrophoresis to strips of nitrocellulose membrane. These strips were incubated with radioactive 18S or 28S rRNA so that they hybridized to the rDNA fragments. The strips were then washed and autoradiographed. The autoradiograph visualized only those bands of DNA containing rRNA coding regions. (cf. the Southern transfer technique section 4.13). In the case of human DNA digested with *Eco* RI, 28S rRNA hybridizes only to a fragment of 8 kb; 18S rRNA hybridizes to the same fragment and to a fragment of about 7 kb as well as varying amounts of longer fragments around 23 kb. The latter hybridization probably reflects the presence of a heterogeneous spacer. 28S rRNA also hybridized to one fragment about 18 kb, of human DNA digested with *Hind* III and 18S rRNA hybridized to the same fragment and to a fragment of 15 kb. This gives a minimum repeat length of 33 kb. These data have been interpreted in terms of a map of human rDNA like that in Fig. 4.32 which is probably generally correct but it is based on a number of assumptions that have still to be checked. This will probably be done by detailed analysis of cloned rDNA (4.7) and by electron microscopy of isolated rDNA as in the case of *Drosophila*.

Such analysis has been carried out for mouse rDNA giving a 44 kb repeat which is heterogeneous. 5.8S rRNA lies between the 18S and 28S rRNAs as in *Drosophila* and *Xenopus*.

It seems likely that tandem repeats of rRNA coding sequences occur in most eukaryotes including very primitive ones like yeast and *Dictyostelium discoideum*. These repeats are integrated into chromosomes and clustered in the nucleolar organizer regions. Exceptions to this arrangement occur in *Physarum* where the rDNA is on extrachromosomal palindromes (see 4.12) and in the amplified

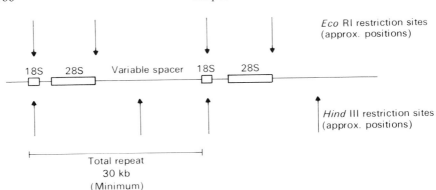

Fig. 4.32 Organization of DNA sequences coding for ribosomal RNA in humans.

rDNA of the *Tetrahymena* macronucleus (also extrachromosomal palindromes) or the amplified rDNA of *Xenopus* and other oocytes (linear tandem arrays). In the normal arrangement the 18S, 5.8S and 28S genes are linked in that order with short transcribed spacer DNA between them and a longer, largely non-transcribed, spacer DNA between one 18S–28S sequence and the next. In *E. coli* this spacer contains other genes, for example tRNA genes, but in eukaryotes the function of the long spacer is unknown. The length of the spacer is heterogeneous even within one species and the average lengths do not correlate with genetic complexity since the primitive cellular slime mould *Dictyostelium discoideum* has a long repeat, 38 kb, comparable with the mammalian repeat, while *Xenopus* has a shorter repeat, 12–17 kb. Similarly the spacer in the two extrachromosomal palindromes described is quite different. In electron micrographs of transcribing rDNA, transcription is not normally detectable in the spacer regions and isolated rDNA does not hybridize to tRNA or 5S rRNA. It is impossible to rule out the presence of other coding sequences but there is no evidence for them in the rDNA spacer sequences in eukaryotes. It remains to speculate that the spacer might have a structural role or a role concerned with the replication and sequence conservation of the coding sequences. In this context it is interesting that spacer sequences apparently evolve much more rapidly than coding sequences, at least in *Xenopus*, and that, again in *Xenopus*, a substantial amount of the length heterogeneity is due to variation between individuals (known as polymorphism) and the heterogeneity within a single individual is quite limited. These results imply that the selection pressure on non-transcribed spacer sequences is low and hence suggest that these sequences have a non-specific function.

4.12 EXTRA CHROMOSOMAL PALINDROMES

Throughout the life cycle of the true slime mould, *Physarum polycephalum*, the genes for 28S, 19S and 5.8S ribosomal RNA are found on linear extrachromosomal DNA molecules (rDNA). There are 100–200 rDNA molecules per nucleus, in the nucleolus. Each rDNA molecule is 65 kb long and the sequence is a *palindrome* which means that it is the same starting from either end, like the word 'tenet' for example. In the case of DNA this property requires that the two strands be identical and that the sequence of half of one strand is exactly complementary to the sequence of the other half of the same strand. This latter feature can be seen directly by observing the renaturation kinetics, which are first order, i.e. independent of the DNA concentration, because each single strand can renature with itself by forming a 'snap-back' or 'hair pin' structure Fig. 4.10. These 'snap-back' structures can be seen by electron microscopy. The palindrome structure is confirmed by the map of restriction enzyme recognition sites (Fig. 4.33) and by the location of the sequences that code for rRNA (also in Fig. 4.33). The locations were determined by hybridizing rRNA to restriction fragments of the rDNA. An electron micrograph of transcribing rDNA has been obtained that shows transcription of the end sequences but not of the central region which thus constitutes a very large 'spacer' region of apparently non-transcribed DNA about 30 kb long.

Unlike the spacers in the other genes studied the overall base composition of the non-transcribed sequences of *Physarum* rDNA (46% dA + dT) is very close to the overall base composition of the transcribed sequences (44% dA + dT). The replication of the rDNA occurs bidirectly from an origin at or near the centre of the molecule, i.e. in the non-transcribed region. The function of the rest of the non-transcribed rDNA is unknown although it may play a structural role or be involved in the conservation of the sequence of the whole rDNA.

The genes for ribosomal RNA are also in extrachromosomal palindromes in *Stylonychia* and *Tetrahymena pyriformis* although in these cases the structure is

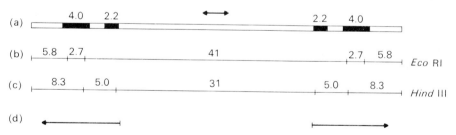

Fig. 4.33 Organization of DNA sequences coding for ribosomal RNA in *Physarum polycephalum.* (a) shows a single, extra-chromosomal, DNA molecule with the 26s (4.0 kb) and 19s (2.2 kb) rRNA coding sequences. The sequence coding for 5.8s rRNA is not shown but lies between the 26s and 19s sequences in the short spacer. Replication of this molecule is initiated near the centre as indicated by the double headed arrow. (b) and (c) show the sites of attack by restriction nucleases *Eco* RI and *Hind* III respectively. The arrows, (d), indicate the sequences that are transcribed and the direction of transcription.

found exclusively in the macronucleus and not at other stages of the life cycle. The *Tetrahymena* rDNA has the same general arrangement as the *Physarum* rDNA except for the size of the non-transcribed spacer which is much smaller in *Tetrahymena*. A *Tetrahymena* cell normally contains both a micronucleus and a macronucleus. Transcription takes place in the macronucleus but only the micronucleus goes through the sexual cycle and maintains genetic continuity: in each cycle a new macronucleus is derived from the micronucleus. The extra-chromosomal rDNA probably occurs only in the macronucleus. The micro-nucleus contains just one copy of half the rDNA palindrome integrated into the remainder of the chromosome and so during the construction of the macro-nucleus the integrated half rDNA must be replicated to form a complete palindrome and then excised from the chromosome to undergo amplification, or alternatively the integrated half rDNA must be replicated, the new half released from the chromosome and then duplicated to form the complete palindrome. These models are illustrated in Fig. 4.34 but it is not known which, if either of them is correct.

4.13 IMMUNOGLOBULIN GENES AND SOMATIC REARRANGEMENT

Immunoglobulins are proteins that are produced by differentiated plasma cells in response to foreign matter in the blood. The immunoglobulins (antibodies) react specifically with the foreign matter (antigens) leading to its rejection. There is a very large number of different immunoglobulin molecules and any particular antibody is made from two molecules of each of two such immunoglobulins. Immunoglobulins can be studied by taking advantage of a type of cancer, myeloma, in which an immunoglobulin-secreting cell proliferates uncontrollably and can be isolated from blood plasma and grown in culture. Such cells continue to synthesize specific immunoglobulins and are a good source of material that can be purified and sequenced.

As an example of an immunoglobulin consider a group of polypeptides, the λ chains. One such λ chain occurs in a specific antibody. A number of λ chains have been sequenced and the general features illustrated in Fig. 4.35 have emerged.

The V region contains most of the variability between different λ chains. It is preceded by the leader region, L, and joined by the J region to the constant, C, region. The C region has the same amino acid sequence in all the different λ chains. Much more is known about the molecular biology of antibodies but we now move on to their molecular genetics.

Immunoglobulin genes, like immunoglobulins themselves, can be studied using myeloma cells. Myeloma cells can be used to obtain immunoglobulin mRNA which is used as a probe for immunoglobulin genes either directly or

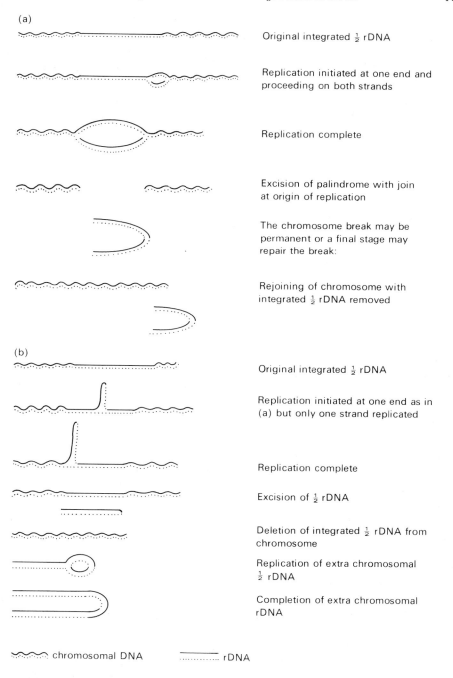

(a)

Original integrated ½ rDNA

Replication initiated at one end and proceeding on both strands

Replication complete

Excision of palindrome with join at origin of replication

The chromosome break may be permanent or a final stage may repair the break:

Rejoining of chromosome with integrated ½ rDNA removed

(b)

Original integrated ½ rDNA

Replication initiated at one end as in (a) but only one strand replicated

Replication complete

Excision of ½ rDNA

Deletion of integrated ½ rDNA from chromosome

Replication of extra chromosomal ½ rDNA

Completion of extra chromosomal rDNA

〜〜〜〜 chromosomal DNA   ⎯⎯⎯⎯ rDNA

Fig. 4.34 Models for formation of extrachromosomal rDNA in *Tetrahymena*.

Fig. 4.35 The λ polypeptide chain (part of an immunoglobulin molecule) showing the different regions of the amino acid sequence.

after cloning and amplifying the cDNA (section 4.7). These probes and a technique known as Southern transfer or gel blotting have demonstrated that somatic rearrangements occur during development of a λ chain-producing cell from the embryo. To see how this rearrangement was discovered we must understand in more detail than before (section 4.9) the Southern transfer technique which is described below, as we go through the experiments showing rearrangement. The total DNA of a particular cell type, e.g. an embryo cell or a λ chainproducing cell, was isolated and digested with a restriction enzyme, e.g. *Eco* RI. The resulting mixture of DNA fragments was fractionated by agarose gel electrophoresis to give a continuous spectrum of fragments of different sizes representing all the sequences in the cell DNA. *Eco* RI cuts at specific points in the DNA so any specific sequence, such as the immunoglobulin gene, will occur on a specific fragment or fragments which will form a sharp band on the agarose gel. For the Southern transfer technique the DNA fragments were transferred from the agarose gel to a nitrocellulose membrane by placing the membrane on the gel and passing a buffer solution through the gel and the membrane. The solution carried the DNA out of the gel and into the membrane where the DNA bound to the membrane. The flow of buffer was arranged by sandwiching the gel and membrane between wet paper (underneath) and dry paper (on top). The dry paper sucked the buffer through (Fig. 4.36). The DNA was denatured and fixed to the membrane by heating at 80°C and then incubated at 68°C in a solution containing the radioactive mRNA or cDNA probe.

The probe will hybridize to DNA fragments bound to the membrane and form bound double strands where the bound DNA fragment has a sequence complementary to the sequence of the probe. In the case of the embryo cell DNA the sequences complementary to λ chain mRNA occur on two fragments, of sizes 3.5 kb and 8.6 kb so bound double strands were formed at the positions on the membrane where DNA fragments of these sizes occurred. The unbound probe was washed off and the membrane examined by autoradiography. The

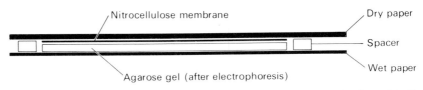

Fig. 4.36 Principle of Southern transfer technique (or gel blotting). The flow of buffer by capilliary action from wet filter paper to dry filter paper carries DNA fragments out of the agarose gel onto the nitrocellulose filter. Modified nitrocellulose filters can be used to transfer RNA or proteins. (See Southern E. M. (1975) *J. Mol. Biol.* **98**, 503.)

autoradiograph showed two bands, at 3.5 kb and 8.6 kb where the radioactive probe had hybridized to the cellular DNA fragments fixed on the filter.

In the case of the λ chain-producing cell DNA the autoradiograph showed three bands, two were the same as the embryo bands, 3.5 kb and 8.6 kb, and the third was a new band at 7.4 kb. The band at 7.4 kb represented a DNA sequence present in the differentiated cell but absent in the embryo cell. Further experiments showed that the 3.5 kb fragment from embryo cell DNA contained part of the λ chain gene (the amino terminus region containing the variability) and that the 8.6 kb fragment from embryo cells contained the rest of the λ chain gene. The 7.4 kb fragment, not found in embryo cell DNA, contained the whole λ chain gene. During differentiation the two parts of the λ chain gene become joined together to produce a single DNA sequence that could be transcribed into RNA for production of the λ chain mRNA. The fact that there are separate parts of the λ chain-producing gene also present in the λ cell may mean that the somatic rearrangement only occurs on one of the two sets of chromosomes in these diploid cells (see discussion on allelic exclusion in chapter 7). The initial and final stages of the somatic rearrangement are shown diagrammatically in Fig. 4.37 which is based on a paper published in *Cell* **15**, 1–14 in 1978 by Tonegawa's group. The diagram shows the DNA sequence as a solid or broken line where it does not code for amino acids and as a box where it does code for amino acids. These coding regions are labelled L, V, J and C to correlate with the different regions of the λ chain polypeptide shown in Fig. 4.35. There are two intervening sequences, I_1 and I_2, which are probably transcribed but removed after transcription to produce the mRNA. The two fragments shown

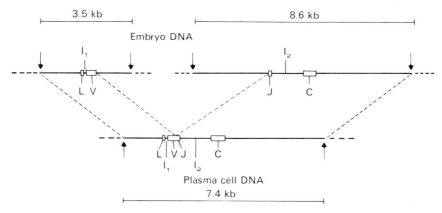

Fig. 4.37 Rearrangement of DNA sequences coding for immunoglobulin. The arrows indicate sites of attack by *Eco* RI. The solid line is non-coding DNA, the hollow line is coding DNA; and the regions L, V, J and C correspond to the polypeptide regions in Fig. 4.35. I_1 and I_2 are non-coding sequences inserted within the C-region coding sequence. (Adapted from Brack C., Hirama M., Lenhard-Schuller R. & Tonegawa S. (1978) *Cell* **15**, 10.)

in the top of the figure are the 3.5 kb and 8.6 kb fragments produced by *Eco* RI digestion of embryo cell DNA. They contain the separated parts of the λ chain gene including the non-coding spacers I_1 and I_2. The fragment shown in the lower part of the figure represents the 7.4 kb fragment found in λ chain—producing cell DNA but not in embryo cell DNA. The DNA sequence in this fragment was generated during differentiation by joining the two parts of the λ chain gene directly within the V + J coding region to generate a single sequence containing the whole λ chain gene.

4.14 INTERVENING SEQUENCES

Firstly, a note about terminology. The DNA sequences that actually code for protein are called 'coding DNA'. The same term is sometimes used for DNA sequences that produce the stable rRNA and tRNA molecules. The DNA sequences that are never transcribed are called 'spacer DNA'. This term is also used for DNA between genes even though it may be transcribed. Within a single length of transcribed DNA there are three types of sequence: coding DNA; sequences which are preserved in the mature mRNA but do not code for protein; sequences which are lost during maturation of the mRNA. The latter sequences have been referred to as 'intervening sequences' above. They are also known as 'introns' or 'inserted sequences'. The first two sequences have been called 'exons'. The term 'split gene' is used for a gene containing intervening sequences.

Intervening sequences were mentioned above in the model of the λ chain gene, in *Drosophila* ribosomal RNA genes and in adenovirus. They have also been shown to occur in immunoglobulin α chain genes, hen ovalbumin genes, chick ovomucoid genes, rabbit and mouse β-globin genes, yeast tRNA genes and *Physarum* ribosomal genes (see Table 7.3). Intervening sequences do not occur within the coding regions for at least one set of sea urchin histone genes nor for some polypeptide hormone genes in mammals. For practical experimental reasons the genes known to have intervening sequences are all rather specialized but it does appear that intervening sequences within the coding regions of DNA are a major feature of the organization of eukaryotic DNA.

Ovalbumin provides an excellent example of intervening sequences in a gene coding for a protein. The organization of this gene has been determined in great detail using molecular cloning, restriction enzyme mapping, nucleotide sequence determination and R-loop mapping. Figure 4.38 shows a map of the ovalbumin gene. The solid line shows the DNA sequence where it is non-coding and the boxes show the sequences that appear in the mRNA, most of which are translated into the ovalbumin polypeptide chain. Four sites of *Eco* RI digestion are shown. The three fragments, a, b and c, containing coding DNA, can be detected in an *Eco* RI digest of total chicken DNA using hybridization with a

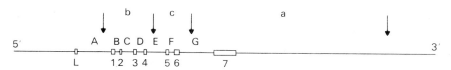

Fig. 4.38 Region of chicken DNA containing the sequence coding for ovalbumin. L, 1, 2, 3, 4, 5, 6, 7 are DNA sequences (exons) that are found in the mature mRNA. L and the 3' end of 7 are not translated. The remainder of the mRNA is translated into ovalbumin. A B C D E F G are DNA sequences (introns) that are transcribed but removed during processing of the transcript to mRNA. The arrows indicate the positions of *Eco* RI attack that are discussed in the text (there are also two more *Eco* RI sites in A). a, b, c are the hybridizable fragments produced by digestion of the DNA by *Eco* RI.

cloned cDNA made from ovalbumin mRNA. The pattern of hybridizeable bands is the same for chick embryo cell DNA, chick erythrocyte DNA and chick oviduct DNA which suggests that somatic rearrangement does not occur with ovalbumin. However, Fig. 4.38 shows that all the *Eco* RI sites are in non-coding DNA so the cDNA prepared from ovalbumin mRNA is not cut by *Eco* RI. This discrepancy, no cuts in the cDNA but at least four cuts in the gene, provided the first clear evidence for intervening sequences in a gene coding for a protein. The length of DNA that is transcribed is probably at least 7.7 kb and it includes all the sequences that appear in the mRNA and all the intervening sequences A . . . G. The intervening sequences, about three-quarters of the transcript, are removed during RNA processing to leave the leader sequence, L, and the coding sequences 1 . . . 7 as the mRNA which is then transported to the cytoplasm for translation.

4.15 EUKARYOTIC 'PROMOTER' SEQUENCES

Table 4.2 shows the DNA sequence preceding the 'cAp' site of ten different nonrepetitive genes which are probably transcribed by RNA polymerase B (II). The 'cAp' site is the position of the 5'-end of the mRNA which has a special end group added after transcription. The 'cAp site' may be the site of initiation of transcription. The sequences in Fig. 4.22 can be compared to see if there are any common features. One such feature is the A–T rich pentanucleotide found about 25–30 bp 'upstream' (i.e. on the 5' side) of the 'cAp' site. This sequence is also found in bacterial promotor regions and is involved in the interaction of RNA polymerase with the DNA. A DNA double helix has about 10.4 base-pairs per turn so the A–T rich pentanucleotide sequence is on the same side of the double helix as the 'cAp site' but 3 turns away.

Table 4.2 can also be used to see the evolutionary changes between mouse α-globin and mouse β-globin which diverged about 500 million years ago and between the two mouse β-globins which diverged about 50 million years ago.

Table 4.2 Possible Eukaryotic Promoter Sequences (Polymerase B)

B. mori fibroin	...TCAGTATAAAAAGGTTCAACTTTTTCAAATCAGC	AT...
Adenovirus II major late protein	...GGCTATAAAAGGGGGTGGGGGCGGTTCGTCCTC	AC...
Mouse immunoglobulin λ light chain	...AGTTATATTATGTCTGTCTCACTGCCTGCTGCTG	AC...
Rabbit β globin	...GGGCATAAAAGGCAGAGCAGGGCAGCTGCTGCTT	AC...
Mouse β globin major	...AGCATATAAGGTGAGGTAGGATCAGTTGCTCCTC	AC...
Mouse β globin minor	...TGGGTATAAAGCTGAGCAGGCTCAGTTGCTTCTT	AC...
Mouse α globin	...GGCATATAAGTGCTACTTGCTGCAGGTCCAAGAC	AC...
Chicken ovomucoid	...TTTGTATATATTTGCAGGCAGCCTCGGGGGGACC	AT...
Chicken ovalbumin	...GCTATATATTCCCCAGGGCTCAGCCAGTGTCTGT	AC...
Sea urchin H2A	...GGTATAAATAGCCAGCAAAAAGATAGGTGGTCA	AC...

(From Portman R., Schaffner W. & Birnstiel M. (1977) *Nature* **264**, 32.)

In this chapter we have seen how the introduction of new techniques has revealed an unexpected variety in molecular genetics of eukaryotes that is being explored with great excitement in laboratories across the world.

4.16 FURTHER READING

Adams R. L. P., Burdon R. H., Campbell A. M. & Smellie R. M. S. (eds) (1976) *Davidson's the Biochemistry of the Nucleic Acids*, 8th edn. Academic Press, New York.

Axel R., Maniatis T. & Fox C. F., (eds) (1979) *Eukaryotic Gene Regulation*. Academic Press, New York.

Bostok C. J. & Sumner A. T. (1978) *The Eukaryotic Chromosome*. North Holland Publishing Company, Amsterdam.

Brack C., Hirama H., Lenhard-Schuller R. & Tonegawa S. (1978) A complete immunoglobulin gene is created by somatic recombination. *Cell* **15**, 1–14.

Bradbury E. M. & Javaherian K. (eds) (1977) *The Organisation and Expression of the Eukaryotic Genome*. Academic Press, London.

Busch H. (ed.) (various) *The Cell Nucleus, Vols. 1, 6, 7*. Academic Press, New York.

Dion A. S. (ed.) (1979) *Structure and Function of DNA, Chromatin and Chromosomes*. Year Book Medical Publishers, Chicago.

Lewin B. (1974) *Gene Expression—Eukaryotic Chromosomes*. John Wiley & Sons, London.

National Academy of Sciences (1977) *Research with Recombinant DNA*. National Academy of Sciences, Washington D.C.

Nicolini C. (ed.) (1979) *Chromatin Structure and Function*. Plenum Press, New York.

Pellegrini N., Manning J. & Davidson N. (1977) Sequence arrangement of the DNA of *Drosophila melanogaster*. *Cell* **10**, 213–224.

Portugal F. H. & Cohen J. S. (1977) *A Century of DNA*. The MIT Press, Cambridge, Massachusetts.

Vogel H. J. (ed.) (1977) *Nucleic Acid—Protein Recognition*. Academic Press, New York.

Watson J. D. (1970) *The Molecular Biology of the Gene*. W. A. Benjamin, New York.

Yunis J. J. (ed.) (1977) *Molecular Structure of Human Chromosomes*. Academic Press, New York.

Chapter 5
The Cell Cycle and Replication

The emphasis in the first four chapters has been on the structure of DNA and chromatin and on the transcription of DNA sequences to produce RNA sequences. The approach has been mainly molecular and we shall be returning to these topics for a more biological viewpoint in the last two chapters. In this chapter we shall study the other main function of DNA, namely replication. The growth and multiplication of cells is inextricably linked with the replication of DNA and so we shall be exploring both the mechanisms of DNA replication and the cell cycle processes with which DNA replication and chromosome structure are integrated.

5.1 CELL PROLIFERATION: THE MITOTIC CYCLE

The mitotic cycle is the normal process that cells use to proliferate or multiply. There are two main events that occur during each cycle: DNA synthesis or replication; division of the parent cell into two offspring. During the mitotic cycle these two major events usually occur at different times and with gaps or growth phases between, Fig. 5.1 shows such a cycle.

Cell proliferation really begins with DNA replication. The period of DNA replication is known as the S phase. ('S' for synthesis is rather misleading since DNA synthesis, for example to repair damaged DNA, can occur at any time; it is the complete replication of the chromosomes that occurs in the defined period known as S phase). During this period each strand of DNA in each chromosome is replicated just once and the new molecules combine with both old and newly synthesized histones and non-histone proteins to make two new chromatids from each original chromosome, (section 5.5 goes into more detail). These processes occur while the chromosomes are dispersed throughout the volume of the nucleus.

After S phase the cell enters G2 phase. During G2 phase the cell prepares for mitosis and cytokinesis. Few of the molecular events that occur in G2 phase have been identified but it is likely that contractile proteins accumulate in preparation for chromosome separation and cytokinesis and that changes in chromosomal proteins, for example phosphorylation of some amino acid residues in H1 histone, occur and cause the initiation of chromosome condensation that leads the cell into mitosis.

The main stages of mitosis are prophase, metaphase, anaphase and telophase followed by reconstruction. Anaphase is normally also followed by cytokinesis.

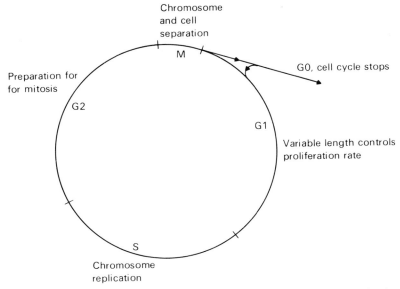

Fig. 5.1 Stages of the mitotic cycle. M, G1, S and G2 are the stages of the mitotic cycle. G0 represents cells that are not going through a division cycle. As an example, in cultured mouse fibroblasts G1, S and G2 occupy 6, 8 and 5 h respectively.

Figure 5.2 shows the stages in diagrammatic form. These stages are clearly recognized in the light microscope and appear to be almost universal among eukaryotes. During prophase each chromosome condenses into a very compact structure with two fat arms, joined by a narrow constriction called the centromere. At the same time the nucleolus and, eventually, the nuclear membrane disappear. The condensed chromosomes line up (i.e. the pairs that were formed by replication of single chromosomes in S phase) on the 'metaphase plate' while spindle fibres form, leading from the centromeres of the chromosomes to opposite sides or 'poles' of the cell. Note that the two homologous copies of the somatic chromosomes do not associate at the spindle in mitosis. This is metaphase. During anaphase the chromosomes move out to the pole regions so that each half of the cell contains exactly one copy of each chromosome. Each set of chromosomes now acquires a nuclear membrane and disperses within the nuclear membrane and the nucleolus reforms. These processes are termed telophase and reconstruction. Now cytokinesis can occur and the cell membrane pinches off where the metaphase plate was and two new cells are formed. Cytokinesis is really the end of the proliferation cycle and some eukaryote cells immediately enter S phase again to continue proliferation at the maximum rate. More commonly, however, the cells enter either G0 phase or G1 phase. In G0 phase proliferation stops, sometimes for ever, but in other cases the cells can return to G1 phase on receipt of a chemical signal that allows proliferation to proceed. Sometimes this signal is the removal of an inhibitor of proliferation.

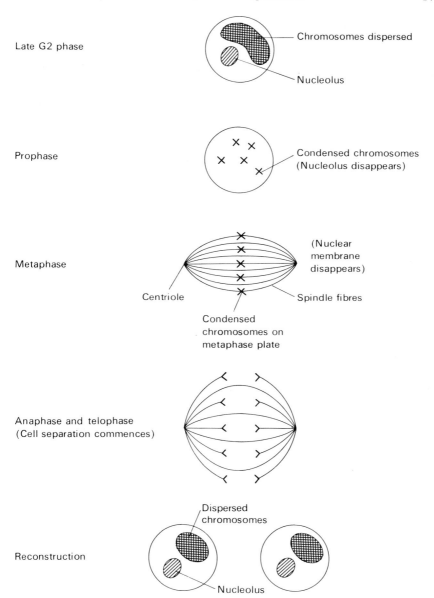

Fig. 5.2 Stages of mitosis.

Such inhibitors are usually specific for a group of cell types and they are called chalones. The important molecular events that occur during G1 phase are largely unknown. Some preparation for S phase occurs but a separate G1 phase is not essential for this. More important are two general processes: differentiation and control of the rate of proliferation. Differentiation of cell types during growth and development involves complex molecular changes to chromosomes and other control systems in the nucleus. These changes probably occur during G1 phase. The rate of proliferation in eukaryotic cells depends primarily on the length of G1 phase. The actual proliferation process of S phase, G2 phase, mitosis and cytokinesis takes approximately 9 h in many eukaryote cell types but G1 phase varies from zero to many hours.

The molecular basis of control of length of G1 phase is unknown. There are several theories about the type of process. The simplest theory proposes that once a cell enters G1 phase there is a fixed probability that a 'certain event' will occur within the cell. When that event occurs it triggers the start signal for cell proliferation and the cell enters the final stages of G1 from where it is committed to one mitotic cycle. An example of the appropriate type of event would be the formation *or* decay of a unique molecular structure. The average length of G1 phase, on this theory, is determined by the probability of the 'certain event' occurring and this probability is different in cells with different growth rates. There are, however, other theories and because of their importance in relation to cancer, much current research is being carried out in this area.

5.2 EVIDENCE FOR SPECIFIC CELL CYCLE STAGES

The evidence generally boils down to a demonstration that S phase is restricted in time and separate from mitosis. The simplest (and oldest) experiment is to take a growing cell culture containing cells at all stages and label for a short period with methyl-^{3}H-thymidine. Cells in S-phase during the labelling period will incorporate ^{3}H into their DNA. A sample of the culture is taken for autoradiography as shown in Fig. 5.3 (describes autoradiography) and the remaining culture is washed to remove unused methyl-^{3}H-thymidine and allowed to grow in non-radioactive medium. Samples are taken at intervals. In the first sample some cells will have incorporated ^{3}H-thymidine, as seen by black spots over the nuclei on the autoradiograph. These are S phase cells. Other cells will be in mitosis as seen by their condensed chromosomes and a third group of cells is neither in S phase nor in mitosis. These are in G1 or G2 phase. If there are N cells on the slide including N_s in S phase and N_M in mitosis then the length of S phase is

$$\frac{N_s}{N} \times \text{ time for 1 cell cycle}$$

and the length of mitosis is

$$\frac{N_M}{N} \times \text{ time for 1 cell cycle.}$$

The time for one cell cycle is obtained as the time for the number of cells in the

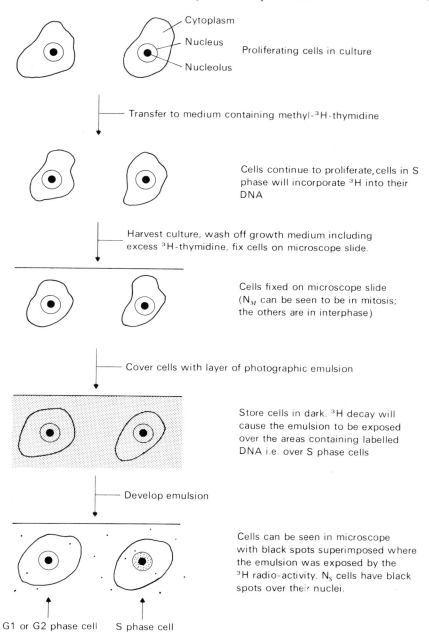

Cytoplasm

Nucleus

Nucleolus

Proliferating cells in culture

Transfer to medium containing methyl-³H-thymidine

Cells continue to proliferate, cells in S phase will incorporate ³H into their DNA

Harvest culture, wash off growth medium including excess ³H-thymidine, fix cells on microscope slide.

Cells fixed on microscope slide (N_M can be seen to be in mitosis; the others are in interphase)

Cover cells with layer of photographic emulsion

Store cells in dark. ³H decay will cause the emulsion to be exposed over the areas containing labelled DNA i.e. over S phase cells

Develop emulsion

Cells can be seen in microscope with black spots superimposed where the emulsion was exposed by the ³H radio-activity. N_S cells have black spots over their nuclei.

G1 or G2 phase cell   S phase cell

Fig. 5.3 Determination of cell cycle stage by autoradiography.

growing culture to double. G2 phase is obtained by studying autoradiographs of successively later samples until labelled mitotic cells are observed. The time between adding the label and obtaining the first labelled mitotic cells is the length of G2 phase. The cycle time can be found as the time required for all the cells to be labelled. An alternative approach is to use synchronously growing cells where all the cells are at the same stage at all times. The best synchrony occurs naturally as in *Physarum polycephalum* where a large flat plasmodium will form containing 10^8 nuclei in a common, well-mixed, cytoplasm. Since nuclear division appears to be controlled via the cytoplasm, all these 10^8 nuclei divide simultaneously and continue to do so as long as nutrients are available. *Physarum* provides an excellent model system for studies of the cell cycle in a eukaryote.

A useful degree of synchrony can be obtained in other systems by selection or induction. Selection involves removing from a culture a group of cells which are at a certain stage of the cell cycle and allowing growth to continue from that stage. For example, during mitosis, mammalian cells change their surface properties and become rather round and smooth so that they are easily shaken off the surface on which they are growing. Mitotic cells can be removed by gentle washing of a monolayer culture and the detached cells collected and allowed to continue growing. Such a culture is initially highly synchronous but soon loses its synchrony through variations in the length of G1 phase from one cell to another. Cells can also be selected for size using centrifugation. Induction involves blocking the cells in an initially non-synchronous culture at a specific point so that they all stop at that point, then when the block is removed the culture starts off in synchrony. Unfortunately, the synchrony again decays rapidly. Induction methods are widely used but are prone to produce artefacts due to the use of blocking agents (these are usually drugs such as colcemid which prevents spindle formation and blocks cells at or near metaphase and hydroxyurea which inhibits DNA synthesis and hence blocks cells at the beginning of S phase. *Tetrahymena pyriformis* can be synchronized by heat shocks which block assembly of the feeding organ).

Measurement of DNA content and cell number in synchronized cell cultures gives data like Fig. 5.3 which shows that DNA synthesis occurs at a specific time different from cytokinesis and allows calculation of the lengths of the cell cycle phases.

More recently, the 'cell sorter' has been used for cell cycle analysis. This instrument measures DNA content of a large number of individual cells.

The DNA is labelled cytologically with a specific fluorescent label and labelled cells from a non-synchronous culture are then passed singly through a laser fluorimeter which measures and records the DNA content of each single cell. Three classes are found corresponding to G1 phase, the minimum DNA content; S phase, intermediate DNA content, and G2 phase, the maximum DNA content of twice the G1 phase amount. The numbers of cells in each phase are proportional to the lengths of the phases. This instrument will sort out the cells in each phase and collect them separately but at the present stage of development the numbers of cells obtained are small.

5.3 MEIOSIS

Eukaryotic cells in G1 phase contain two copies of each chromosome and are therefore said to be diploid. The copies are normally genetically different and are referred to as homologues. Cells involved in reproduction, namely the gametic egg and sperm cells, are haploid, containing only one homologue of each type of chromosome. The fusion of the two gametes, one from each parent, then recovers the normal diploid chromosome complement. Haploid cells are produced from diploid cells by the process of meiosis—the stages are shown in Fig. 5.4. Meiosis involves two rounds of cell division. In the first round, the cell, having passed through S phase, has pairs of homologous chromosomes each consisting of twin sister chromatids. When these chromosomes condense, the homologues move alongside one another and fuse to form a bivalent of homologous chromosomes, often called a tetrad. These bivalents then line up on the spindle. (Note that this association of homologous chromosomes is unique to meiosis—at mitosis no association of homologues occurs and each chromosome attaches to the spindle independently.) Prior to the first cell division the bivalent breaks and one of each homologue moves to each pole. This is therefore a *reduction* division. This is followed by a second meiotic division in which the chromosomes again condense, but this time the behaviour at the spindle is as in mitosis, and division and separation of the constituent chromatids provide the genetic endowment for the new cells, the gametes, each containing only one copy of one version of each homologous chromosome. During the first meiotic division, exchange of DNA also occurs between the four chromatids of the bivalent tetrad (see Fig. 5.5). This process is known as 'crossing over' and gives rise to genetic recombination.

5.4 GENETIC RECOMBINATION

Very early in the history of cytogenetics, microscopists observed that meiotic chromosomes, at stages between pachytene and the first metaphase, displayed curious patterns of chromosomal cross-over. These cross-overs were seen to involve two chromatids of the paired homologous chromosomes associated within a tetrad (from pachytene onwards the meiotic chromosomes can be found to exist as tetrads of paired homologous chromosomes, each chromosome itself comprising two sister chromatids). Some long chromosomes were often found to be involved in two or three separate chiasmata, and, as meiosis proceeded and the homologues were pulled apart on the spindle to the spindle poles,

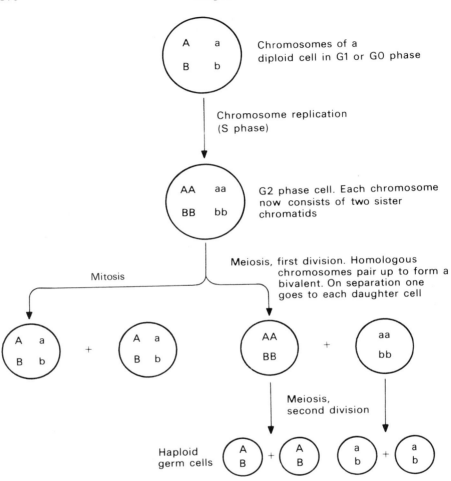

Fig. 5.4 Mitosis and meiosis. Each letter A, a, B, b . . . D, d represents one
chromosome A and a represent non-identical versions of one chromosome type; B and b
represent non-identical versions of a different chromosome type. The sex chromosomes
are not shown on this diagram. In the G2 cell the double letters represent the twin
chromatids of one chromosome, held together at the centromere.

the chiasmata appeared to terminalize (see Fig. 5.6), permitting entire separation
of the constituent chromosomes.

It is now clear that such terminalization occurs only with a few chiasmata
near to the ends of relatively short chromosomes. More frequently a second
break and rejoin in the DNA occurs, permitting the normal outcome of recombination namely a reciprocal exchange of homologous lengths of DNA.
The way in which the first cross-over of the DNA comes to be followed by the
second cross-over, thus permitting the cross tracking of entire lengths of DNA,

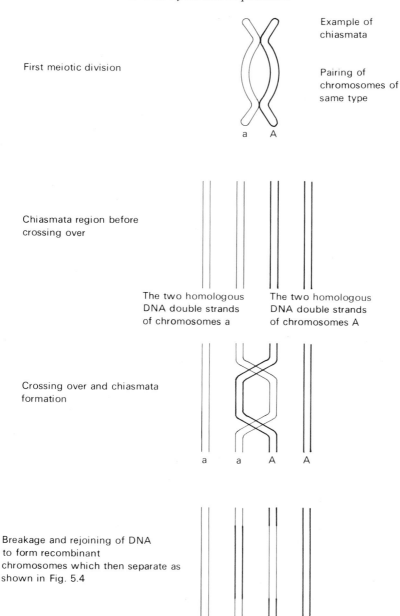

First meiotic division

Example of chiasmata

Pairing of chromosomes of same type

Chiasmata region before crossing over

The two homologous DNA double strands of chromosomes a

The two homologous DNA double strands of chromosomes A

Crossing over and chiasmata formation

Breakage and rejoining of DNA to form recombinant chromosomes which then separate as shown in Fig. 5.4

Fig. 5.5   Crossing over.

is probably accomplished by a process outlined in theory by Dr Robin Holliday, and frequently termed the Holliday Model of Recombination (see Fig. 5.7). This model suggests that the second break in the DNA occurs at a narrow bridge

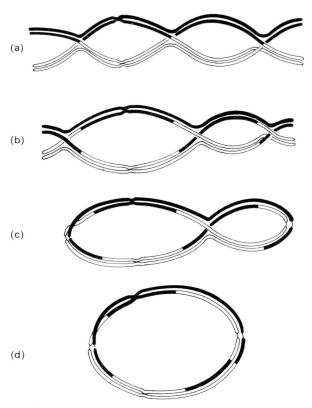

Fig. 5.6 Diagram to show the presumed sequence of terminalization of chiasmata. The constructions indicate the centromeres, and the shading the parental origin.

(a)–(d) Successive stages from early diplotene to diakinesis, showing how three interstitial chiasmata become two terminal chiasmata. The points where crossing-over has taken place remain unchanged. (From Whitehouse E. L. K. (1973) *The Mechanism of Heredity*. Edward Arnold, London.)

which results from the realignment of chromatids involved in a recombination event. This theoretical concept is now supported by microscopical and bio-chemical evidence.

Genetic analysis by linkage studies and by radiolabelling or other marking procedures (Figs 5.8 and 5.9) convincingly demonstrate that crossing-over and chiasma formation involve breakage and reunion of the DNA strands on the chromosomes in the chiasma, so that DNA which was previously part of one chromosome comes to be linearly connected to another. Crossing over, therefore, accomplishes an exchange of homologous DNA between chromosomes, and genes which were previously known to be on one homologue can be found to have been transferred to the other.

At one time it was assumed that the breakage and rejoining of the DNA

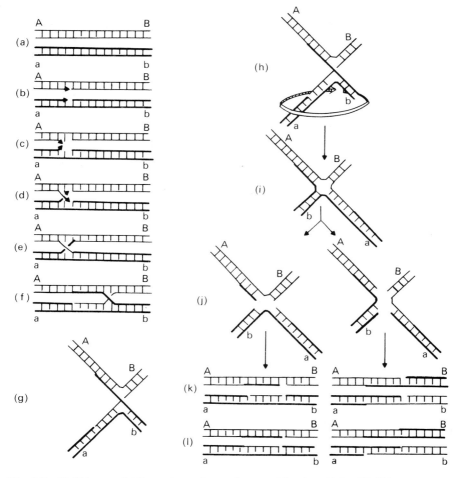

Fig. 5.7 Holliday model for reciprocal genetic recombination. (After Holliday R. (1974) *Genetics* **78**, 273 with permission.)

followed from chiasma formation, but there is now good evidence that the appearance of these chiasmata follows from the actual DNA recombinational event (this used to be known as the chiasma type theory of recombination and was originally proposed and supported by such geneticists as Janssens and Darlington).

Although many factors, both genetic and environmental, have been found to affect crossing-over and the incidence of chiasma, the mechanism at the level of the DNA remains obscure. A major cytological structure is known to be involved, termed the synaptonemal complex (see Fig. 5.10), and it comprises a protein rich zone of association between homologous chromosomes at early pachytene. Although many interesting theories have been advanced to explain

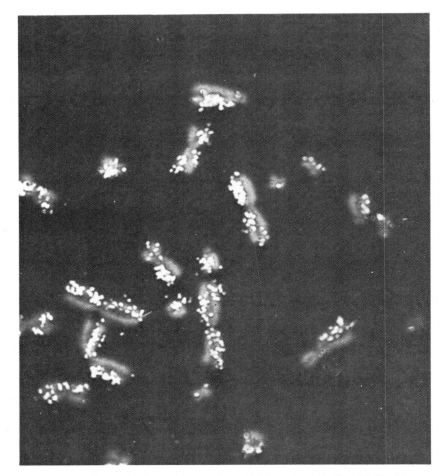

Fig. 5.8 Autoradiograph of second metaphase division after labelling DNA. All chromosomes possess one labelled chromatid. Sister chromatid exchanges are responsible for a switch of label between some of the two chromatids. Note that in this figure and in Fig. 5.9 what is represented is exchange between sister chromatids prior to mitosis. Meiosis is not involved (see Latt *et al.* (1980)). (Photograph kindly provided by Prof. David Prescott, with permission from Lewin B. (1974) *Gene Expression Vol. 1.* John Wiley & Sons, Chichester.)

the mechanism of DNA exchange within the complex (see Fig. 5.11), little of the evidence is hard.

We should not leave the phenomenon of crossing over without making two further points about it. The first is that, however it is effected, it constitutes a highly important mechanism of genetic recombination which therefore accomplishes genetic reassortment and increased variation in eukaryotes. Without crossing over, gene mutation would persist but gene reassortment would be

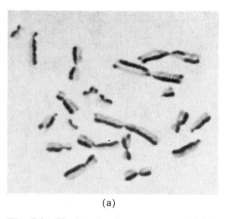 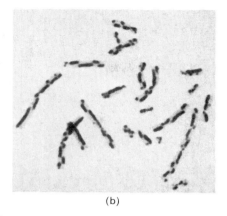

(a)   (b)

Fig. 5.9 Harlequin chromosomes. (a) Metaphase cell from Chinese hamster ovary cell line following two rounds of replication in 5 bromouracil followed by Hocchst/ Giemsa staining. Only one chromatid of each sister-chromatid pair is stained. (b) As in (a) but exposed to nitrogen mustard (HN_2) for two cell cycles before sampling. The rate of sister chromatid exchanges is approximately 10 times higher than in (a). (Photograph kindly supplied by Prof. H. J. Evans. (From Perry P. & Evans H. J. (1975) *Nature* **258**, 121.)

minimized (except for the more or less random assortment which seems to be a common feature of organisms with large chromosome numbers). Such assortment may not involve actual exchange and is not truly subject to genetic control.

The other point is that recombination between separate DNA molecules is, of course, a common phenomenon in prokaryotes, as exemplified by experiments in bacterial conjugation, transformation and transduction, (see Chapter 3 for fuller discussion of these topics). Whether the interesting insertion sequences (1S elements), found to be involved in some situations of genetic recombination in micro-organisms, also function in eukaryotic crossing over is not at present known (see page 265). It seems, however, a probable hypothesis. Recombination may also be evident between sister chromatids at mitosis, especially following U.V. irradiation (see Figs. 5.8 & 5.9).

5.5 SYNTHESIS OF CHROMOSOMAL COMPONENTS IN THE CELL CYCLE

DNA synthesis is mostly confined to S phase except for repair of damaged DNA, synthesis of DNA in cytoplasmic organelles, (for example mitochondria and chloroplasts) and synthesis of extrachromosomal DNA in the nucleus, (for example viral DNA and the genes for ribosomal RNA in *Physarum polycephalum*.) During S phase DNA synthesis occurs at many sites simultaneously. For the purposes of replication DNA may be regarded as made up from separate

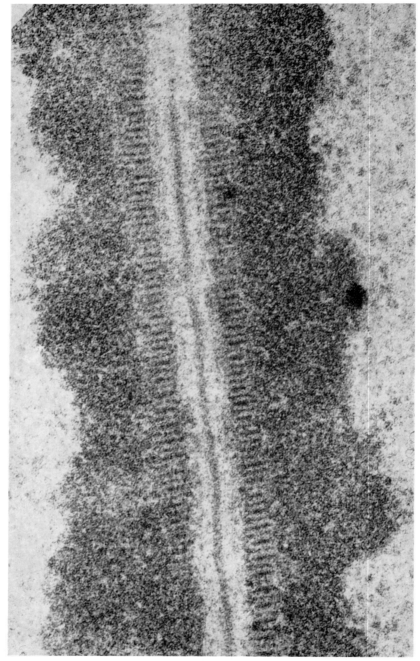

Fig. 5.10   Electron micrograph of synaptonemal complex. (Photograph kindly supplied by Prof. M. Westergaard.)

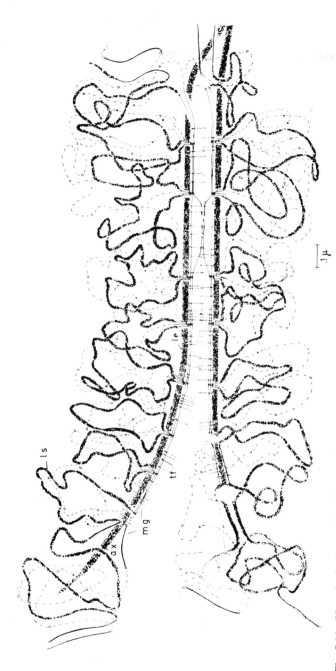

Fig. 5.11 Hypothetical assembly of a synaptic complex from two axial cores. The central element in this model assumed to include DNA spun out from the two homologues on either side. (From DuPraw E. J. (1970) *DNA and Chromosomes*. Holt Rinehart and Winston Inc. New York. (After Moens P. B. (1968) *Chromosoma* **23**, 498.))

sequences, called replicons, joined together to make a chromosome DNA. Each replicon has an origin of replication where replication of the replicon begins. A loop is formed by the two new double strands of DNA, as replication proceeds. The ends of the loops, called the replication forks, move away from the origin causing the replication loop to grow until the whole replicon is replicated. The movement of both replication forks away from each other can be seen by pulse labelling cells with ^{3}H-thymidine and then examining the replicating DNA molecules by electron microscope autoradiography. The ends of the loops are both labelled as shown diagrammatically in Fig. 5.12 showing that replication proceeds in both directions simultaneously i.e. bidirectionally.

The initiation of replication of particular replicons occurs in a particular order during S phase. In *Physarum polycephalum* the very good synchrony allows the following experiment to be carried out very precisely. During a short period of one S-phase the DNA is labelled with radioactive thymidine so that only DNA synthesized at the particular part of S-phase is labelled. Then, in the following S-phase, a density label, bromodeoxyuridine, is added for the equivalent period. The cells are harvested and the DNA analysed on a density gradient which separates the DNA labelled with bromodeoxyuridine from the normal density DNA. The radioactive label coincides with the bromodeoxyuridine label in this case only when equivalent labelling periods are used for the two labels, showing that the DNA sequences being replicated at a particular time in one S phase are replicated at the same time in the subsequent S phase. A similar situation has been demonstrated in HeLa cells. It has also been found that, in mammalian cells, some strongly heterochromatic chromosome regions are replicated at the end of S phase. The initiation of replication of a replicon

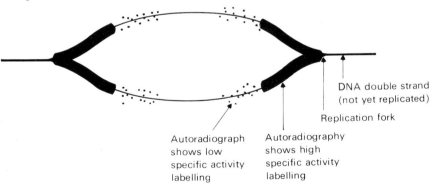

DNA double strand
(not yet replicated)

Replication fork

Autoradiograph
shows low
specific activity
labelling

Autoradiography
shows high
specific activity
labelling

Fig. 5.12 Autoradiography of replicating DNA. This is a similar experiment to that of Fig. 5.3 but at much higher resolution and the cells were labelled first by growing for a short time in a medium containing ^{3}H-thymidine at low specific activity followed by a short time in medium containing ^{1}H-thymidine at high specific activity. DNA synthesized in the first period is lightly labelled with ^{3}H and DNA synthesized in the second period is heavily labelled. The results show the general direction in which DNA synthesis is proceeding.

occurs at a specific point called the origin of replication. Initiation may involve the interaction of a specific protein with a specific DNA sequence but much remains to be discovered concerning the initiation of DNA replication. Once initiation is complete the two replication forks move apart and replication proceeds.

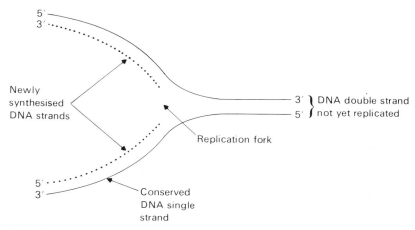

Fig. 5.13 Semi-conservative replication of DNA. The replication fork is moving from left to right. Each new DNA double strand has one single strand from the original parent DNA and one new single strand.

Consider the movement of one replication fork, called fork migration. Fork migration requires the synthesis of two DNA double strands to replace the one original double strand. This occurs semi-conservatively i.e. each new double strand includes one of the original single strands, (Fig. 5.13). Semi-conservative replication was established experimentally by Meselsohn and Stahl using density labelling techniques.

Meselsohn and Stahl grew *E. coli* cells in a medium containing the 'heavy' isotope, ^{15}N, of nitrogen so that the DNA incorporated ^{15}N. This made the DNA denser than normal and it could be separated from normal DNA, containing only ^{14}N, by centrifugation to equilibrium in a density gradient of CsCl. *E. coli* cells containing ^{15}N–DNA were transferred to normal medium containing ^{14}N so that all subsequent DNA synthesis would use only ^{14}N. After one round of DNA replication only one type of DNA was seen by density gradient analysis. Its density was half-way between that of ^{15}N and that of ^{14}N–DNA consistent with each new double strand having one old strand (^{15}N) and one new strand (^{14}N). After two rounds of replication, half the DNA had normal density (^{14}N–DNA) and half the DNA had the same density as after one round of replication. Together, these results prove that DNA is replicated semi-conservatively. The experiment is summarized in Fig. 5.14.

Three enzymes capable of extending a DNA chain are known, polymerase I,

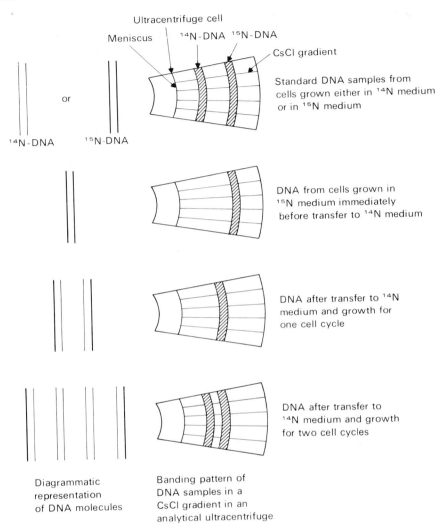

Fig. 5.14 Density gradient analysis of DNA. Density labelling experiments show that DNA replication occurs by a semi-conservative mechanism.

polymerase II and polymerase III. Polymerase I is probably used *in vivo* to repair damaged DNA. It finds widespread use *in vitro* for copying or labelling DNA molecules as mentioned in the previous chapter. Polymerase III is probably the enzyme involved in replication. These enzymes extend DNA chains in the 3'–5' direction by adding a nucleoside 5' phosphate residue onto the 3'–OH of the growing chain. (The nucleoside 5' phosphate residue is derived from a nucleoside 5' triphosphate). Careful inspection of Fig. 5.13 will reveal a dilemma since the polymerase can easily 'copy' one strand of the template DNA as the replication fork migrates but the other strand points in the wrong direction for direct 'copying'. It turns out that DNA synthesis occurs discontinuously and the new strand is synthesised in short segments of a few hundred nucleotides called 'Okazaki fragments'. Each 'Okazaki fragment' is started with a short RNA strand. This is later replaced with DNA. The 'Okazaki fragments' are joined up as replication proceeds by the enzyme DNA ligase. Both new strands use this mechanism. Migration of the replication fork also requires 'unwinding' proteins that reduce the stability of the original DNA double helix. The replication fork is clearly a rather complex structure and it is not yet possible to reproduce it *in vitro*. Figure 5.15 summarizes, in diagrammatic form, this outline of DNA replication. The synthesis of histones is normally co-ordinated with the synthesis of DNA although the co-ordination can be

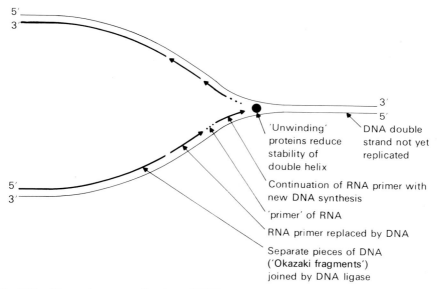

Fig. 5.15 Discontinuous replication of DNA.

broken by inhibitors. The control of histone synthesis appears to be exerted at the point where the RNA produced by transcription of the histone genes is processed to yield cytoplasmic histone mRNA. Transcription of histone genes appears to occur throughout the cell cycle but histone mRNA only accumulates during S phase. Consequently, histone synthesis only occurs during S phase.

As a replication fork moves along the chromosome, the nucleosome structure is at least partly disassembled and reassembled again. This is shown by the high sensitivity to nuclease of newly synthesised DNA. However, the assembly of nucleosomes is probably not random. Several experiments have shown that

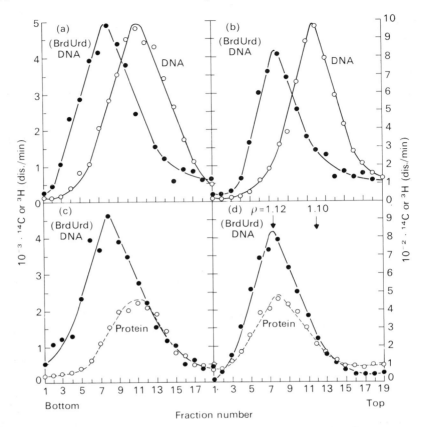

Fig. 5.16 Distribution of histones on new and old DNA strands after chromosome replication. (a) and (b) show two standard gradients, each with density labelled DNA strands ([BrdUrd]DNA) and normal DNA strands separated by centrifugation. The [BrdUrd]DNA is newly synthesized. (c) shows the [BrdUrd]DNA profile again (—●—) together with the profile of radioactivity in old histone. (d) shows the [BrdUrd]DNA yet again, this time with the profile of radioactivity in new histone. Clearly, the new histone bands with the new DNA strands and the old histone with the old DNA strands. (From Russev G. & Tsanev R. (1979) *Eur. J. Biochem* **193**, 127.)

newly synthesised histones are bound to the newly synthesised DNA strand while the original histones remain bound to the original DNA strand. Figure 5.16 shows the results of one such experiment. Each panel shows the result of centrifuging DNA-histone complexes on a density gradient at pH 11.5 where the DNA strands separate. The histones had been fixed onto their associated DNA strand by chemical cross-linking. In (a)–(d) the solid lines represent newly synthesised DNA which is denser because it contains bromodeoxyuridine instead of some thymidine. In (a) and (b) the fainter lines are the original DNA which is of normal density. In (c) and (d) the broken lines represent protein; (c) shows the original protein; (d) shows newly synthesised protein. The results show that new histones are associated with the new DNA strand and old histones are associated with the old DNA strand.

Transcription does not stop during S phase. In fact, in *Physarum* plasmodia most mRNA is made during S phase. Transcription does, however, stop during mitosis when the chromatin is condensed. The structure of transcriptionally active chromatin is different from that of inactive chromatin (section 5.9) and this structure is presumably consistent with the structural requirements for S phase but inconsistent with the structure of the metaphase chromosome.

5.6 CHANGES IN CHROMOSOME STRUCTURE DURING THE CELL CYCLE

During mitosis chromosomes form very compact well defined structures that can be seen and characterized by light microscopy. In some eukaryotes, for example *Drosophila*, the chromosomes differ from one another morphologically and the complete set of chromosomes can be readily described and significant differences between each chromosome of the set recognized. Such a description is called a karyotype. In organisms with more chromosomes, such as mammalian cells, the karyotype is determined partly by gross morphology and partly by using staining techniques, for example Giemsa staining, that produce characteristic banding patterns on the chromosomes allowing them to be readily distinguished from one another (see section 2.9 in Chapter 2). The molecular basis of the banding patterns is currently under study. The chromosome structure appears, from nuclease digestion experiments, to contain nucleosome sub-units and the possible ways in which these are arranged in the metaphase chromosome have been discussed in Chapter 2.

During the rest of the cell cycle, called interphase, the chromosomes are dispersed throughout the nucleus and individual chromosomes can no longer be recognized. The chromosomes, usually called chromatin when dispersed, are largely invisible in the light microscope but some dense regions can be seen, particularly by electron microscopy. These two forms are known as euchromatin and heterochromatin respectively. During interphase there are progressive changes in overall chromatin arrangement. Some hints that this is occurring can be obtained from electron microscopy and from studies of the binding of drugs or dyes, such as acridine orange, to DNA in nuclei at different cell cycle stages. More dramatic changes can be seen by fusing cells at different stages of interphase with cells at metaphase. This is accomplished experimentally, using inactivated Sendai virus to make cells 'sticky', and when fusion between a metaphase and an interphase cell occurs, the chromatin of the interphase nucleus undergoes a dramatic change known as premature chromosome condensation. If the interphase nucleus is in G2 phase the prematurely condensed chromosomes (sometimes just called P.P.C.) look very similar morphologically to metaphase chromosomes although they are not so compact. If the interphase nucleus is in S phase the pattern of prematurely condensed chromosomes is much more disordered but it is possible to distinguish regions containing replicated chromatin (like G2 phase prematurely condensed chromosomes); regions containing non-replicated chromatin (like late G1 phase prematurely condensed chromosomes); and regions that have not condensed properly which are assumed to be sites where replication is in progress. If the interphase nucleus is in G1 phase the prematurely condensed chromosomes may be highly condensed (early G1 phase) through several stages to relatively extended (late G1 phase). In fact, the morphology of the prematurely condensed chromosomes can be used to determine the stage of G1 phase. The pattern of prematurely condensed chromosomes in bone marrow cells is being used experimentally to monitor the progress of patients suffering from leukaemia.

These experiments continue to emphasize that the overall chromosome structure is continually changing through the cell cycle.

5.7 CHROMOSOME CHANGES IN DIFFERENTIATION

Generally speaking, different types of cell in one organism contain the same DNA sequences in the same relative amounts. There are a number of exceptions to this generalization, for example amplification of genes for ribosomal RNA during amphibian development or during formation of the macronucleus in *Tetrahymena*; under-replication of simple sequence DNA in polytene chromosomes; rearrangement of DNA sequences during lymphocyte formation and loss of all DNA in mammalian erythrocytes. However, even in typical cells, there is evidence for very many changes in chromosome structure during differentiation. At the level of individual genes, differences in structure are revealed by their sensitivity to pancreatic deoxyribonuclease I (DNase I). The actual DNA sequences probably do not change during differentiation in most cases, (but see Chapter 4) although changes in methylation of cytosine may occur. The changes in structure are probably associated with changes in the relationship between the DNA and chromosomal proteins. For example, histone acetylation appears to be correlated with DNase I sensitivity and a number of workers have found HMG proteins associated with actively transcribing DNA sequences. HMG proteins are non-histone proteins whose properties have been described in Chapter 1. The structure of transcribing chromatin is discussed in more detail in section 5.9.

Superimposed on these differences in structure at the gene level are gross changes in overall chromosome structure that can be detected in certain cell types. These changes usually involve the formation of a highly condensed structure as in avian erythrocytes, in fish sperm and in *Tetrahymena* micronuclei. The change in structure is associated with a change in the very lysine-rich histone H1. In the avian erythrocyte the condensation of the chromatin is accompanied by the progressive displacement of H1 by a new highly basic histone called H5 that is even more effective than H1 at aggregating DNA. In fish sperm phosphorylation of H1 and displacement of H1 by protamine accompany the condensation of the chromatin and in *Tetrahymena* micronuclei the H1 is replaced by an uncharacterized basic protein. In these cases, condensation is accompanied by loss of transcriptional activity. A normal chromatin structure can be recovered in each case; by further differentiation in *Tetrahymena*; by fusion with an egg in the case of fish sperm; and by fusion *in vitro* using Sendai virus and an active cell type in the case of avian erythrocytes. This indicates that the severe condensation observed in these situations remains open to reversal.

5.8 HISTONE MODIFICATIONS

In most types of cell, histone molecules are metabolically stable and the histone: DNA ratio is constant. Consequently, histones are expected to provide the underlying chromosome structure which can be modified in particular regions and at particular times as required for transcription, replication, mitosis or other events. This expectation is borne out by the ability of histones to form the core particle of the nucleosome and by the existence of a nucleosome repeat structure at almost all times and places. Packing of nucleosomes together and modifications to the nucleosome structure itself can be controlled by non-histone proteins, but the histones themselves have a vital role to play through reversible modifications to their structure. The main changes that occur have been described in Chapter 1. There are two particular occasions when modifications have been observed: in certain stages of the cell cycle; during induction of transcription of previously inactive genes.

In the cell cycle two major types of modification have been observed, phosphorylation of H1 and acetylation of H3 and H4. Phosphorylation of H1 has been studied in detail and the sites of phosphorylation associated with growth—serines and threonines in the *N*- and *C*-terminal regions of H1—have been described in Chapter 1. Phosphorylation at serine 37 is discussed below. Many functions for H1 phosphorylation have been proposed and it is entirely possible that this interesting molecule is involved in the entire cycle of chromosome changes already discussed as being part of the whole mitotic cycle. There is a very clear correlation between the amount of H1 phosphorylation and the rate of cell growth in mammalian cell types ranging from those, such as normal liver cells, that are in an almost stationary phase with a very long G1 phase, to those,

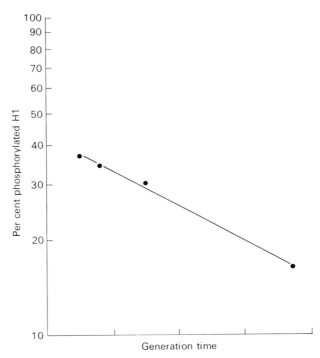

Fig. 5.17 Phosphorylation of histone H1. Normal and transformed liver cells were grown and the histone H1 analyzed. Different cell lines gave different generation times and the graph shows that short generation times gave high levels of phosphate on H1 histone. (From Balhorn R., Morris H. P. & Chalkley R. (1972) *Cancer Research* **32**, 1789.)

such as cancer cells in culture, that are growing very rapidly with a short G1 phase. Figure 5.17 shows a graph of the amount of phosphate bound to H1 as a function of growth rate for different cell types.

Studies with synchronized cells, especially with the naturally synchronous slime mould, *Physarum polycephalum*, have shown that phosphorylation of H1 occurs throughout the cell cycle but that there is a balance between phosphorylation and de-phosphorylation that varies through the cycle. The balance depends on the activities of two enzymes, or groups of enzymes. One, histone kinase, transfers the γ-phosphate of adenosine 5′ triphosphate (ATP) to the growth associated sites on H1 as shown schematically in the reaction:

$$\text{adenosine}-P_i-P_i-P_i + \underset{\text{histone}}{H1} \xrightarrow[\text{Mg}^{2+}]{\text{kinase}} \underset{\substack{\text{phosphorylated}\\\text{histone}}}{H1-P_i} + \text{adenosine}-P_i-P_i$$

and the other, histone phosphatase, removes the phosphate bound to H1 as shown schematically in this reaction:

$$\underset{\substack{\text{phosphorylated}\\\text{histone}}}{H1-P_i} \xrightarrow[\text{Mg}^{2+} \text{ or Ca}^{2+}]{\text{phosphatase}} \underset{\text{histone}}{H1} + P_i$$

During G1 phase the overall effect is to make the H1 phosphate content low, i.e. 0–1 mole phosphate per mole histone. During S phase there is synthesis of new unmodified H1 molecules but the enzymes rapidly bring this up to the same level (1 phosphate per molecule) as the old H1. This phosphate may be involved in assembly of nucleosomes. Much larger changes occur during G2 phase and early mitosis. There is a substantial activation of histone kinase activity in G2 phase which leads to an increase in H1 phosphate content from 1 to 6 phosphates per molecule (Fig. 5.18). The increase in H1 phosphorylation coincides with

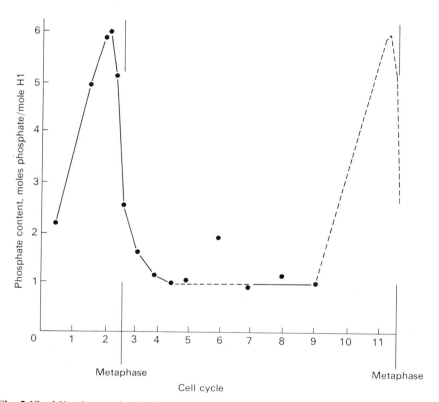

Metaphase   Metaphase

Cell cycle

Fig. 5.18 Mitosis-associated phosphorylation of H1 histone. The number of phosphate groups per molecule of H1 histone is shown as a function of time in the cell cycle for *Physarum polycephalum* plasmodia. The data shows a major increase in H1-phosphate in mitosis as the chromosomes are condensing immediately before metaphase.

the first stages of chromosome condensation in prophase and this and other evidence suggests that phosphorylation of H1 is the main change causing the chromosomes to condense in prophase. In fact, the data suggests that the activation of histone kinase is the major 'triggering' event that, by phosphorylating H1, sets off the final stages of the cell cycle, namely mitosis and cytokinesis.

The mechanism by which phosphorylation of H1 can pack nucleosomes or coils of nucleosomes closer together remains a mystery. *In vitro* studies were

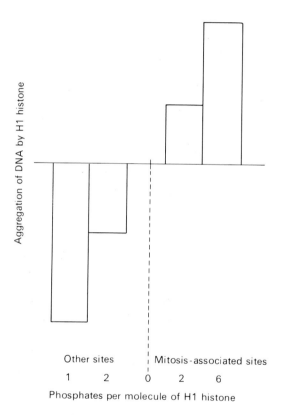

Fig. 5.19 Phosphorylation of H1 histone at specific sites increases its ability to aggregate DNA. The ability of H1 histone to aggregate DNA was deduced from the turbidity of H1–DNA solutions. The diagram shows that if H1 is phosphorylated at sites (ser 37 and/or ser 106) not associated with mitosis the turbidity falls but if H1 is phosphorylated at the mitosis-associated sites the turbidity rises.

originally hampered by the lack of a method for reconstituting chromatin structures containing H1 and by the difficulty of preparing large amounts of the highly phosphorylated form of H1. However, as in Fig. 5.19, studies of the interaction of H1 with DNA showed that phosphorylation of H1 at the growth associated sites increased its ability to aggregate DNA and this increase was specific to the growth associated sites.

Acetylation of H4 also shows major changes in the cell cycle. Figure 5.20 shows the changes in tetra-acetylated H4 during the cell cycle in *Physarum*.

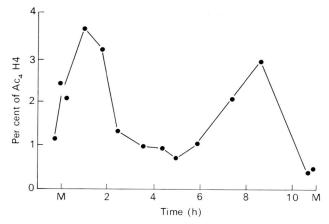

Fig. 5.20 The proportion of histone H4 containing four acetyl–lysine residues as a function of time in the cell cycle. The data shown are for the naturally synchronous plasmodia of *Physarum polycephalum*. The cell cycle is shown from one mitosis (M) to the next. In this particular cycle there is no G1 phase and S phase occupies about 2 h immediately after mitosis, the rest of the cycle being G2 phase.

Early in mitosis there is a minimum amount of tetra-acetylated H4 which correlates with the maximum H1 phosphate content and with chromosome condensation. Hence, the absence of highly acetylated H4 probably produces a 'tight' nucleosome that is suitable for chromosome condensation but unsuitable for transcription which stops during mitosis. The two maxima of highly acetylated H4, one in S phase and one in G2 phase, are probably correlated with transcription which shows a similar double peak during the cell cycle in *Physarum*.

A second major feature of H4 acetylation during the cell cycle is a very rapid turnover of acetate groups of H4 during S phase. This turnover is probably associated with the movement of the DNA replication fork and the disassembly and assembly of nucleosomes that must accompany this movement. An increase in H4 acetate would loosen the nucleosome structure in advance of the replication fork and then removal of H4 acetate would re-tighten the structure onto newly synthesised DNA. Acetylation and subsequent deacetylation of newly synthesised H4 may also be part of the assembly process for new nucleosomes.

Histone H4 acetylation has also been very clearly implicated, particularly by the work of Professor Allfrey and his colleagues, in the transition from an inactive chromatin structure to a structure that is active in transcription. For example, lymphocyte cells can be stimulated by the plant lectin, phytohaemagluttinin. The stimulation causes new transcription to occur which is preceded by histone acetylation. Similar results have been obtained in many other systems. More recently, Dr V. Ingram discovered that the addition of isobutyric acid to growing HeLa cell cultures caused the acetylated forms of H3 and H4 to accumulate. This is due to inhibition of the histone deacetylase. Nucleosomes isolated from isobutyrate treated cells contain acetylated histones and are much more susceptible to digestion by deoxyribonuclease I (DNase I) than normal nucleosomes. This correlates with the increased susceptibility of active nucleosomes. Finally, since the action of DNase I on normal total chromatin is first to digest active DNA sequences, the proteins released from chromatin by limited digestion with DNase-I are those proteins that were associated with the active nucleosomes. Analysis of proteins released by DNase I showed that acetylated histones are indeed associated with the more susceptible chromatin.

All these experiments correlate acetylated forms of H3 and H4 with transcriptionally active chromatin regions.

Finally, phosphorylation of H1 at a specific site, namely serine 37 in the primary sequence, is clearly implicated in the action of cyclic AMP mediated hormones. A group of hormones, for example insulin, acts by binding to receptor sites on the plasma membrane of target cells. This causes activation, inside the plasma membrane, of the enzyme adenyl cyclase. Adenyl cyclase causes the formation of cyclic AMP in the cytoplasm and the cyclic AMP—known as the 'second messenger'—diffuses throughout the cell. One target for this cyclic AMP is the enzyme, protein kinase. Cyclic AMP activates protein kinase by binding to the inactive form of protein kinase and causing it to dissociate and release the active (or 'catalytic') sub-unit. The whole range of *in vivo* substrates for cyclic AMP activated protein kinase is not known. However, Professor T. A. Langan and his colleagues have shown that one effect in liver cells is the phosphorylation of serine 37 on about 2% of the H1 molecules of the cell. Serine 37 is not one of the sites involved in growth associated phosphorylation (see above) so it probably has a separate function, perhaps associated with the changes in chromosome structure that accompany changes in transcriptional activity.

5.9   THE SUB-UNIT STRUCTURE OF TRANSCRIPTIONALLY ACTIVE CHROMATIN

This heading assumes that there is a sub-unit structure in active chromatin, but what evidence is there that this is so? The evidence for a sub-unit structure of chromatin in general is overwhelming, as described in Chapter 2. The main results were obtained by micrococcal nuclease digestion; electron microscopy and neutron diffraction. None of these three techniques allows us to conclude that all (i.e. more than 90%) of the chromatin is in the sub-unit structure. We don't know just how much of the chromatin is being transcribed, or is available for transcription, at any time in a particular type of cell but it could easily be only 10% or less, so the structure of this fraction of chromatin could go undetected by the techniques mentioned above.

How can we study the structure of active chromatin at the molecular level?

(Studies at other levels, microscopic and genetic, are described in Chapter 7). Only by using techniques that differentiate between active and inactive chromatin. Electron microscopy can do this if the chromatin is isolated with the growing RNA chains still attached. Figure 4.3 shows genes for ribosomal RNA being transcribed. In this case the length of the primary RNA transcript is known because it can be isolated and measured and the length of DNA in the transcribing region can be obtained. The 'packing ratio' of the DNA chain in the nucleosome is 6:1, i.e. the length of a chain of nucleosomes is one sixth the length of the DNA molecule present. In Fig. 4.3 the length of the transcribing chromatin is about 0.8 times the length of DNA in the transcribing chromatin. This is about 5 times longer than would be expected if the DNA were packed into nucleosomes showing that transcribing ribosomal genes are not packed into nucleosomes identical with those described in Chapter 2. However, it doesn't rule out the possibility, discussed below, that a modified sub-unit structure may persist even in such an actively transcribing gene.

It is very hard to generalize this result to other active genes. Transcribing ribosomal RNA genes can be recognized in electron micrographs by their localization in the nucleolus, their high density of growing RNA chains, and their reiteration. Other genes cannot be so easily recognized and usually the size of the initial RNA molecule is unknown, so the length comparison cannot be made. It is possible to look for nucleosomes on other active regions in electron micrographs but there is some confusion, due to the fact that nucleosomes and RNA polymerase molecules look alike. Many microscopists confirm the absence of nucleosomes in active regions but this conclusion is always open to the criticism that the structure may have 'stretched out' during preparation for electron microscopy. Note that this criticism cannot be applied to the genes for ribosomal RNA because in their case the polymerase molecules are packed tightly together, which would not be the case if stretching had occurred. In other cases, particularly where the growing RNA chains are well spaced out, nucleosome structures are observed on the DNA between growing chains. However, these may not be truly active.

Active genes can also be studied using the technique of hybridization described in the previous chapter. In this technique, a radioactive probe is prepared with a sequence complementary to the sequence of one of the strands of the active gene. Excess of the probe is then hybridized to a DNA preparation, and the amount of hybridization is a measure of the number of the active gene sequences present in the DNA preparation. When chromatin is digested by micrococcal nuclease and nucleosomes are prepared by sucrose gradient centrifugation the nucleosome DNA can be isolated and hybridized to a probe for active genes. Very substantial hybridization occurs, which indicates that the active genes are in a sub-unit structure. This seems to contradict the electron microscopy evidence, especially since, in some cases at least, the sub-unit repeat length is the same in transcribed and non-transcribed regions, although it does vary from cell type to cell type. However, it may be that the sub-unit itself is different—more extended—although the repeat length remains the same and the bulk of the DNA in the sub-unit remains resistant to micrococcal nuclease. The notion of a different sub-unit is supported by the fact that there are two small differences in the digestion patterns of active and inactive chromatin: (1) for ribosomal genes in *Xenopus* and *Physarum* the more active the tissue the fewer ribosomal genes are found in the nucleosome DNA, which implies that active ribosomal genes are less resistant to micrococcal nuclease than inactive DNA; (2) for ribosomal genes in *Physarum* and ovalbumin genes in chick oviduct, the active genes are digested preferentially to mononucleosomes and occur only to a small extent in chains of nucleosomes.

Much more striking differences between active and inactive chromatin sub-units are seen by digestion with DNase I. As we described in Chapter 2 and mentioned above, DNase I makes single strand breaks within nucleosomes and when the products of digestion are examined on denaturing gels a series of lengths 20, 30, 40 . . . bp is found. These fragments are too short for good hybridization but the undigested DNA can be hybridized readily. Experiments along these lines reveal that the active genes in the many cases now examined are digested first by DNase I, leaving the inactive DNA to be digested more slowly. The same series of fragment lengths is obtained at short digestion times. These results imply that the active chromatin sub-unit is different from the bulk chromatin sub-unit but that some characteristics of the two sub-units remain the same.

DNase I digests the active chromatin sub-unit very fast so it is not a good tool for preparing purified sub-units for further study. Another enzyme, DNase II, is more promising in this respect. The action of DNase II, in the presence of divalent ions, is somewhat similar to that of micrococcal nuclease. DNase II can be used to prepare nucleosomes which can be separated by zone sedimentation on sucrose gradients. The nucleosomes can also be fractionated by solubility in magnesium chloride or sodium chloride. DNA isolated from the soluble nucleosomes contains the active gene sequences. At the time of writing this is a very active area of research.

The differences between active and inactive chromatin sub-units are not due to the presence of different DNA sequences since the same sequences can be either active or inactive depending on the cell type. Therefore the differences must be due to the presence of RNA polymerase or to changes in the protein contents of the sub-units. Data on the protein content of active chromatin sub-units is rather sparse at the time of writing but some possibilities are beginning to emerge. The most fruitful type of experiment to date has used DNase I. Chromatin in nuclei is digested with DNase I for a short period so that principally active chromatin is digested. DNase I releases the proteins bound to the digested chromatin and so these can be collected by centrifuging the undigested chromatin. Analysis of these proteins reveals a substantial enrichment in HMG non-histone proteins particularly HMG 14 and HMG 17 (see Chapter 1). These proteins have also been found in active sub-units prepared using other nucleases and so they may well be involved in modifying the bulk chromatin sub-unit (nucleosome) to form the extended active chromosome sub-unit. There is similar evidence that acetylated forms of the histones H3 and H4 are associated with active chromatin sub-units.

Finally, a clue to the structure of active chromatin sub-units may be obtained from an interesting finding with chromatin from the true slime mould *Physarum polycephalum*. Micrococcal nuclease digestion of *Physarum* nuclei gives the normal nucleosome pattern plus a new type of sub-unit. Most of the active ribosomal genes occur in the new sub-unit. This new sub-unit contains the normal core particle DNA length of 145 bp but its sedimentation cofficient is less than half that of a normal nucleosome showing that the structure is very extended, as required for an active chromatin sub-unit by the electron microscope evidence. The new sub-unit has a modified histone content and substantial amounts of non-histone proteins that resemble HMG proteins. These kinds of experiments will soon give us a much more detailed picture of the structure of active nucleosome sub-units.

5.10 FURTHER READING

Axel R., Maniatis T. & Fox C. F. (eds) (1979) *Eukaryotic Gene Regulation.* Academic Press, New York.

Bostock C. J. & A. T. Sumner (1978) *The Eukaryotic Chromosome.* North Holland Publishing Company, Amsterdam.

Bradbury E. M., Inglis R. J. & Matthews H. R. (1974) Control of cell division by very lysine rich histone (FI) phosphorylation. *Nature* **247**, 257–61.

Busch H. (ed.) (various) *The Cell Nucleus, Vols 2, 3, 4, 5, 6.* Academic Press, New York.

DuPraw E. J. (1970) *DNA and Chromosomes.* Holt, Rinehardt & Winston Inc., New York.

Ford E. H. R. (1973) *Human Chromosomes.* Academic Press, London.

Goldberger R. F. (ed.) (1979) *Gene Expression.* Plenum Press, New York.

Goldstein L. & Prescott D. M. (eds) (1979) The structure and replication of genetic material. In *Cell Biology, Vol. 2.* Academic Press, New York.

Goldstein L. & Prescott D. M. (eds) (1979) Gene expression: the production of RNAs. In *Cell Biology, Vol. 3.* Academic Press, New York.

John P. (ed.) (1980) *The Cell Cycle.* Cambridge University Press, Cambridge, (*in press*).

Kornberg A. (1974) *DNA Synthesis.* W. H. Freeman & Co., San Francisco.

Latt S. A. *et al.* (1980) Sister chromatid exchanges. *Adr. Human Genet.* **10**, 267–332.

Li H. J. & Eckhardt R. A. (eds) (1977) *Chromatin and Chromosome Structure.* Academic Press, New York.

Losick R. & Chamberlin M. (eds) (1976) *RNA Polymerase.* Cold Spring Harbor Laboratory, New York.

Molineux I. & Kohiyama M. (eds) (1977) *DNA Synthesis.* Plenum Press, New York.

Stahl F. W. (1979) Special sites in generalized recombination. *Ann. Rev. Genet.* **13**, 7–24.

Yunis J. J. (ed.) (1977) *Molecular Structure of Human Chromosomes.* Academic Press, New York.

Chapter 6
Transcriptionally Active Chromosomes

Perhaps the greatest impediment to an understanding of chromosome structure and function is that most chromosomes are visible when the chromatin is relatively inactive—at mitosis—and invisible when it is active—during the rest of the cell cycle. To date no one has devised a way of visualizing chromatin in the light microscope during interphase, and electron microscopy of interphase chromatin is difficult and preliminary. Studies of mitotic chromosomes do, of course, provide information about gene arrangement and the packaging of genes in chromatin, but reveal little about gene activity and gene expression.

Fortunately there are two special types of chromosomes which can be visualized microscopically while involved in transcription. It follows that their contribution to our knowledge of chromosome structure and activity has been enormous. These are the lampbrush chromosomes found in some insect tissues and in the primary oocytes (unfertilized eggs) of many vertebrates, especially those of Amphibia, and the giant polytene chromosomes found in some ciliate protozoans and many insect tissues, particularly those of the salivary glands of *Drosophila*. Since transcriptionally active chromosomes are rare and unusual it is well to be cautious about assuming that what holds true for them also holds for all other active chromatin. Moreover, the two types of chromosome are actually very different from one another in many ways, both in their structure and activity, so that if either one of them is taken as the true model chromosome, very widely divergent ideas about general chromosome architecture and function result. It is interesting also to observe that the view of any one geneticist is influenced by the material on which he happens to work, so that *Drosophila* geneticists tend to take the insect polytene chromosome as a model and extropolate from it, while vertebrate and especially developmental geneticists often favour the lampbrush model.

We will proceed to examine each of these fascinating structures in reasonable detail and then move on to a comparison and attempted correlation.

6.1 THE LAMPBRUSH CHROMOSOME

These structures are actually diplotene stage meiotic chromosomes, that is, they are to be found only in the developing oocytes within ovaries. They are known to occur in a wide range of animals, including man, but have been most extensively studied in the oocytes of amphibians, especially newts. The diplotene stage of the first meiotic division is very prolonged, sometimes lasting for periods of 6 months, in these animals, and since they have many large eggs with considerable amounts of DNA in their genomes, the lampbrush chromosomes are startlingly big, sometimes being as much as a millimetre or more in length. Chromosomes with a somewhat similar structure have been observed in a few other special situations, for example in the unicellular green alga *Acetabalaria*, and also in the spermatocytes of some insects, although in this latter example only the Y chromosome adopts the lampbrush configuration. It is presumably largely a result of the extremely extended time scale of meiotic diplotene that transcriptional activity becomes necessary for these chromosomes. After all, a normal mitosis has come and gone within a few hours, amounting to only a short pause in the customary activities of the eukaryotic cell cycle.

The name lampbrush stems from Victorian times. We might more readily term them bottle brush chromosomes, since they resemble the long brushes with which the Victorians cleaned the glasses of their oil lamps and we the insides of our bottles or test tubes. They were actually named by Ruckert in 1892, who discovered them in the oocytes of dogfishes. A glance at Figs 6.1 and 6.2 will reveal that the greatly extended backbone or axis of these chromosomes supports an outgrowth of frills, but that whereas in our bottle brushes the outgrowths are simple bristles or fibres, in the lampbrush chromosome the outgrowths are loops. These loops vary greatly in size and character and are actually in pairs—a point at which the bottlebrush analogy breaks down and ceases to be useful.

6.1.1   The structure of lampbrush chromosomes

By way of beginning our discussion of lampbrush chromosomes, we will do well to briefly mention how they are most readily observed and studied. Unlike mitosis, in which the nuclear envelope normally disintegrates and the chromosomes occupy a large part of the cell, the nuclear membrane persists in meiosis and therefore must be broken if a clear view of these chromosomes is to be obtained. The technique now commonly employed to study these chromosomes involves puncturing an oocyte of the appropriate stage, and recovering the intact nucleus with a Pasteur pipette after it has been squeezed from the cell. Following transfer of the nucleus to a special glass chamber, having a microscopic coverslip as its *base*, the nucleus is broken either by mechanical or chemical means and the long chromosomes permitted to flatten down in the aqueous medium onto the coverslip. They can then be viewed and studied by light phase microscopy, but with the objective lens below and not above the cover slip. Turning the phase microscope upside down in order to view such structures from below was a simple but important step forward in their study. When viewed in this way such chromosomes are unfixed and presumed to be identical in form to those within the diplotene oocyte. Actually they are often somewhat obscured by protein and other nuclear material in such preparations,

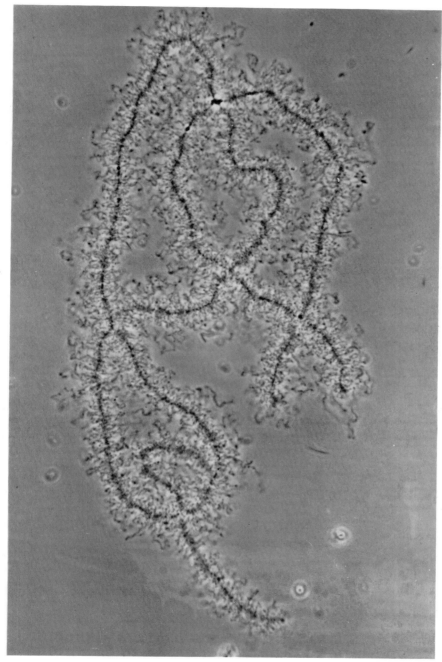

Fig. 6.1 A bivalent lampbrush chromosome from the newt *Triturus viridescens,* photographed by phase contrast. X476. (Courtesy of Prof. J. G. Gall.)

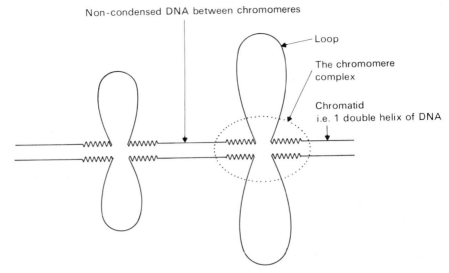

Fig. 6.2   Organization of DNA in the lampbrush chromosome.

and may require either subsequent teasing out or gentle enzymic digestion with proteolytic enzymes to reveal details of their structure.

A close study of the main axis of these chromosomes reveals that it consists of a succession of bead-like structures, the so called chromomeres, and in some favourable places these may be seen to be separate, implying that they are joined together by an exceedingly thin threadlike structure. The chromomeres are slightly variable in size, being from 0.25 to 2 μm in diameter. Since, as was stated earlier, an entire chromosome can be as much as one millimetre in length when stretched, it follows that there may be up to 2000 of these chromomeres packed along one chromosome. The true number is probably something less than this, depending on the species in question; in any event counting the chromomeres is by no means easy. Each chromomere can be seen to support one or more pairs of loops, and although these loops vary somewhat in size and shape, the two loops of a pair are identical to one another. Each loop is actually somewhat asymmetrical, but we will return to a discussion of that point later. Just as the chromosome number varies from species to species, so must the total number of loop pairs, but in Amphibia the overall number of chromomere loop pair units varies from 5000 to 20 000 for the entire chromosome set.

So far it has been established that a single lampbrush chromosome consists of a string of beads, the chromomeres, and to each bead is attached two identical loops. We should now ask more precisely what the relationship is between the chromomeres and the loops, and indeed the thin axial thread joining the chromomeric beads together. If, by gentle micromanipulation, a single lampbrush chromosome were pulled out and stretched lengthwise, it might be expected that the intrachromomere threads would break. But actually, as was first observed by Professor Callan of St Andrews University, who has pioneered studies on these chromosomes, when pulled out, the beads themselves come apart, so that the loops come to constitute part of the central axis. Since these chromosomes are at the diplotene stage of meiosis, near to the end of the first meiotic division (see section 5.3), each chromosome assembly will consist of two homologous chromosomes held together at any chiasmata which may have formed (see Fig. 6.1). But, in addition, each homologue of the pair will itself be double, consisting of two sister chromatids. Viewed in this way, the paired loops would represent homologous loops on two sister chromatids. It might, therefore, be supposed that each chromomere, apparently single, would in fact be double, and the connecting axial thread, too fine to be visualized by light microscopy in any event, to be double also. These predictions have now been borne out by good experimental evidence, some of it carried out in Callan's laboratory at St Andrews, but much of it by Gall at Yale. For example, digestion of lampbrush chromosomes with DNase enzyme leads to breakage of the loop material much more rapidly than breakage of the axis. Such a result strongly supports the idea that the loops consist of one DNA double helix while the axis consists of two double helices. It can therefore be concluded that, as indicated in Fig. 6.2 each chromatid corresponds to one very long DNA molecule. This DNA is, however, in three different arrangements, being either extended as a short linker between chromomeres, extended as a long loop from one chromomere, or highly condensed in the chromomere. In addition, we also know that each chromomere can itself be broken into four parts, two from one chromatid and two from the other, the two parts of one chromatid being, in fact, the opposite ends of a single DNA loop. To be cautious, it should be stated that the equivalence between one loop and one half chromomere, although probable, may not be invariably correct. So too with the stretching of one chromomere to force the loop to form a bridge—it may be that sometimes the chromomere breaks at one end so leaving, not two small chromatin aggregates, but only one, the loop leading straight into the linker DNA at one end.

We have actually jumped the gun here a little in naming DNA as the axis and loop fibre material. But such is undoubtedly the case. Although both protease and RNase enzymes digest away much of the material attached to the loops, they leave, apparently intact, the entire fibre backbone of the chromosome. Such experiments also indicate that, if protein linkers are inserted anywhere in the axis of the chromosome, then they are protected in some rather special way to render them highly resistant to protease enzymic digestion. On the other hand, digestion with DNase enzyme is highly destructive to both loops and main axis and both are eventually disrupted entirely by such digestion. Measurements in the electron microscope help to confirm such conclusions about the nature of loops and axis. Following trypsin digestion, the loop axis measures only about 2.5 nm in diameter, the dimensions of a single DNA double helix. Similarly the main axis was shown to be double and to be made up from a backbone of two DNA double helices running in parallel.

Having confirmed the DNA nature of the chromosome, it is now reasonable to move on to ask how much DNA is allocated to the various components of the chromosome, loops, chromomeres and connecting axis. Since each loop is only about 50 μm long and a haploid set of lampbrush chromosomes may contain, say, 4000 loops (this varies greatly with species, of course, as does the overall DNA content) a total loop length of 20 cm is likely. But since such an amphibian will, on the basis of its haploid DNA content, yield not less than 500 cm of DNA in total, it can be seen that most of this DNA must be packed up in the chromomeres. Only a little of the total DNA is needed to provide the linkers stretched between chromomeres, certainly less than 0.5 cm. It follows that, of the total DNA, some 5% is in the loops, 0.1% in the stretched main axis, and say 95% in the aggregated chromomeres. To recapitulate, if a single chromatid were stretched out without breakage, leaving the chromomere material condensed, progress from one end to another would meet a bit of straight linker, then an aggregate which, in strict terms is one quarter of a chromomere, then a long stretch of non-condensed loop DNA, then another aggregate, another linker and so on.

Although the loops of the lampbrush chromosome account for only 5% of its total DNA, they are in many respects its most important parts, since they appear to be the sites of transcription. There is no evidence for RNA synthesis taking place within the chromomeres or even on the uncondensed linker DNA in the axis, although at least in the latter, the entry of polymerase molecules does not seem to be entirely excluded on grounds of physical access, as it is from the chromomeres. So we can address ourselves to a closer look at the loops and what goes on there.

6.1.2   A Closer Look at the Loops

The first observation to be made following even a superficial glance at a set of lampbrush chromosomes, is that not all the loops are the same. Firstly there are some loop pairs which, at least in certain species, are noticeably larger than others (see Fig. 6.3). These include a pair of loops on chromosome XII of the newt *Triturus* which Gall and Callan identified and termed giant granular loops. Such loops are not only long but are very densely enshrouded with a granular matrix of ribonucleoprotein particles. So too, a pair of loops in the salamander *Plethodon cinereus*, although not so densely covered as the former loops, are astonishingly long, extending for about 240 μm from end to end.

Secondly, members of a pair of homologous loops on partner chromosomes may sometimes be seen to differ in morphological appearance. Now since such a pair of loops constitute homologous portions of two distinct chromosomes, we might expect that, at least in a situation of genetic heterozygosity, the loops would differ, and since the morphology of the loops is actually an accumulation

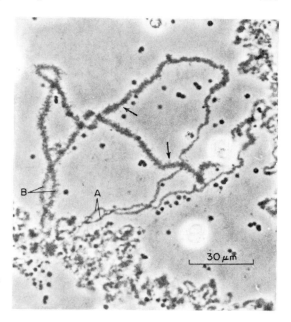

Fig. 6.3 Phase contrast photomicrogroth of a lampbrush bivalent from the salamander *Plethodon cinereus* showing a pair of very long loops, both arising from the same chromomere. Each loop has a thin insertion (a) and a thick end (b). (From Macgregor H. C. (1977) In *Chromatin and Chromosome Structure*. Li H. J. & Eckhardt R. (eds) Academic Press, New York.)

of transcriptional product on the loop DNA, that is a mass of RNA and protein, this genetic difference might conceivably be transmitted as a morphological difference between one loop and its opposite number. Such seems to be the case; indeed in some examples, this differing morphology can be followed in related animals and seems to segregate as a simple Mendelian factor. This all goes to reinforce the original comment about the great value of being able to visualize transcriptional activity in these nuclei.

Having compared differing loops, it should now be asked whether a single loop is constant in its morphological appearance from one end to the other. The answer is that, in many cases at least, it is not. All loops are enshrouded by a rather bulky matrix of RNA and protein—otherwise they would be invisible in the light microscope—but the distribution of this matrix is not uniform, tending to be thin at one end of a loop and thick at the other. The actual significance of this observation we will return to shortly.

6.1.3 Transcriptional Activity on the Lampbrush

As has been emphasized, the loops of the lampbrush chromosome are visible in the light microscope as a result of their transcriptional activity. The transcription of the DNA leads to an accumulation of RNA *in situ*, some of it no doubt still in process of completion, other molecules finished but still present within the chromosomal matrix. The intensity of RNA synthesis on the loops, as revealed by autoradiographic studies using tritiated uridine, suggests that, at least on some loops, the RNA polymerase molecules must be following one another fairly closely round the loop. But we have not yet established whether the whole loop is transcribed, and if so, whether as one complete unit or a series of units. It has become quite clear that, whereas in some loops transcription goes continuously from start to finish round a loop leading to a very large RNA molecule, in other loops there are two or more separate transcriptional units. As shown in Fig. 6.4 up to five separate transcribing units may exist on some loops, and these may not all run in the same direction. This of course implies that the two strands of the double helix are in use, but probably not overlapping parts of the same section, although this cannot be ruled out. Some very beautiful electron microscopy of lampbrush loops has revealed that the RNA molecules become larger as the polymerase molecules traverse (see Fig. 6.5) the loop or loop section, so producing a pattern resembling a Xmas tree (see Fig. 6.5) (and so indeed they are now called in the jargon).

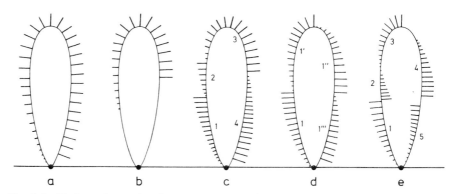

Fig. 6.4 Various alternatives for arrangements of transcriptional units within individual loops of lampbrush chromosomes. Units numbered 1 to 1‴ are presumed to be of equal length, while 1–5 are of different lengths. (Reproduced by kind permission of Dr V. Sheer.)

In addition to the 'Xmas trees' themselves, some apparently untranscribed regions can be detected in some loops.

The size of RNA transcripts from the loop DNA can be determined with reasonable accuracy by scrutinizing the 'Xmas trees' on electron micrographs. They vary from about 5 µm to up to 100 µm, giving molecular weights of about 10^5 daltons to well over 10^6 daltons.

6.1.4 What Kind of RNA is Made on the Loops?
It is of great interest to know what kind of RNA is produced on loop DNA. One thing is certain, and that is that some of it is very large and certainly much larger than most messenger RNA. But it is also clear that much of the RNA synthesized on loops does have a tail of poly-adenine residues, that is, that after its transcription it is modified in a fashion known to be characteristic of mes-

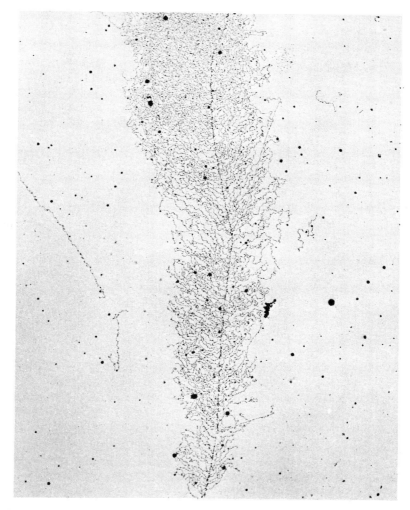

Fig. 6.5 Portion of loop from *Triturus* chromosome, showing 'Xmas tree' formed by the synthesis of RNA, the product increasing in length with distance from the axis. (Courtesy of O. C. Miller & B. R. Beaty.)

senger RNA precursor. Some elegent experiments of Old *et al.* (1977) have involved probing for histone mRNA on lampbrush loops with a radioactive histone cDNA probe. Their photographs reveal that the RNA which hybridizes with the probe constitutes only a part of the entire transcript, and that the histone message seems to be released from the chromosome long before transcription of the whole loop is completed (see Figs 7.11 & 7.12). So it seems reasonable to conclude that most loop RNA is a fraction of the heterogeneous nuclear RNA (itself a probable messenger RNA precursor) which is poly-adenylated and later may be partially degraded to yield messenger RNA. Of course, all sorts of fascinating questions remain unanswered about lampbrush loop RNA. Are all species of mRNA represented in it, and how much of it is multiple copy? We will return to consider these questions shortly.

Although the ribosomal genes are represented on the loops, most of the ribosomal RNA in the oocyte is manufactured by a special process termed gene amplification and this is discussed more fully in Chapter 7. Suffice it to say that numerous copies of the ribosomal RNA genes are produced, actually in an earlier stage of meiosis—pachytene, and these small independent strings of ribosomal RNA genes each produces a mass of rRNA and so comes to resemble a mini nucleolus. The meiotic oocyte nucleus is therefore filled with very many such nucleoli and so only a minute fraction of rRNA synthesis occurs on the chromosomal loops proper. But interestingly, such loops do not seem to be long enough to permit transcription of all the adjacent copies of rRNA present in that part of the genome.

6.1.5   The Protein on the Loops

As was emphasized earlier, the gross morphology of the loops in the light microscope results from the bulky matrix of RNA and protein. Whence the protein? Is it produced in the nucleus on the loop RNA, or is it synthesized in the cytoplasm and transported onto the loops from there? In the main, the latter appears to be correct. Arguments remain in the scientific literature about the validity or otherwise of nuclear protein synthesis, but from the evidence it appears that protein synthesis is the sole prerogative of the cytoplasmic ribosomes. It follows that the mass of protein associated with the loop RNA has been transported from the cytoplasm and must be involved in the formation of ribonucleoprotein (RNP) complexes, some of which are no doubt the sources of messenger RNA which associates with the cytoplasmic ribosomes in translation.

But another question about loop protein remains and that relates to the important issue of nucleosome structure and function discussed in Chapters 1 and 2. Lampbrush chromosome DNA, like all other intra-cellular DNA, is associated with histones to form nucleosomes. How are these nucleosomes distributed on lampbrush chromosomes? On chromomeric DNA, loop DNA, or both? Present evidence is conflicting on the subject, but it seems very likely that nucleosomes are present both on loops and chromomeres, but are simply less tightly aggregated in the former. Part of the difficulty in solving this crucial question arises from the close resemblance in electron micrographs between RNA polymerase enzymes and nucleosomes. Small aggregates of protein on loop DNA which some authors take to be nucleosomes others interpret as RNA polymerase molecules. Certainly these enzymes must be present and in large numbers judging from the close packing of the foliage of the loop 'Xmas trees.' After all, each RNA molecule in the tree must have a polymerase enzyme at its base. The question about the presence or absence of nucleosomes on the loops is closely bound up with ideas about nucleosome function, and, assuming that nucleosomes are present on loops, their spacing certainly implies that H1 histone does not link adjacent nucleosome beads together in the loops as it is assumed to do in much chromatin, including the chromomere complex.

6.1.6   Does the Loop Move?

Since most of the chromosomal DNA is in the chromomere and only 5% in the loops, it is reasonable to ask whether this 5% is constantly transcribed or whether it is on the move, perhaps spun out at one end and spun in at the other. In this way, over the extended time span of diplotene, most or all of the genome could be transcribed. Here again the evidence is, if not conflicting, at least a little complicated. There is little evidence which suggests such movement of loops, and RNA synthesis appears to be equally active all round the loop. But in the case of some very large loops, and especially the so called giant granular loops of the XIIth chromosome of the newt *Triturus cristatus cristatus*, it does look from autoradiographic evidence as if all RNA synthesis occurs at the insertion or thin end of the loop and, once made, the RNA moves round to the thick end before discharge. Whether the DNA moves, or only the RNA molecules still attached to the polymerase, is not absolutely certain. We can only conclude that, in relation to most loops at any rate, there is strong evidence for movement of the RNA and none for movement of the DNA.

THE MASTER AND SLAVES MODEL AND THE LAMPBRUSH CHROMOSOME

A now famous model of gene organization was proposed some years ago by Professor Callan, stemming from his studies on the lampbrush chromosome of newts (see also Chapter 7). Since they, along with so many other organisms, had much more DNA in their genomes than seemed strictly necessary on a one gene–one protein basis, and yet the data from studies on mutation rate suggested just such a relationship, Callan proposed that many copies of most or all genes existed. These multiple copies were, however, in his terminology, slave copies. Alongside them in tandem was one identical 'Master' copy. Correction of slave base sequence to that of the master was proposed as a regular event, so explaining the mutational evidence. It was clearly the possibility of a lamp-brush loop being spun out and gathered in, and only being transcribed when extended, which motivated Callan to propose the theory. Unfortunately there is now a lot of evidence which renders the theory improbable, except perhaps in the case of genes for histone or ribosomal RNA. Hybridization experiments seem to emphasize convincingly that most genes exist as one or very few copies in the haploid genome. So whether or not lampbrush chromosome loops move, it is now clear that commonly there is only one gene per loop and, where more than one does occur, they are often entirely different.

6.1.7   DNA Sequences and Genes on the Loop

One point which deserves mention before leaving these interesting structures is the general phenomenon of DNA sequence interspersion discussed at length in Chapter 4. If structural genes are normally or frequently interspersed with shorter sequences of moderately repetitious DNA, does such interspersion appear in the expression of the lampbrush chromosome? The short answer is that it does. As we have said previously, some loops appear to have more than one gene along their length, but such multiple sequences seem to be all of somewhat similar lengths. But there is evidence for short untranscribed spacer sequences between genes on many of the loops, or at either end of a gene where only one gene occupies a loop. So the lampbrush chromosome studies tend to confirm the interspersion idea, but also suggest that some such moderately repetitious sequences may not normally be transcribed.

6.2   THE POLYTENE CHROMOSOME

A polytene chromosome is one which has replicated many times but retains the replicates within a single structure. Giant polytene chromosomes were first discovered by Balbiani in 1881, who noted that they consisted of a series of dark rings alternating with paler regions. These are now known as bands and interbands. Large banded chromosomes are to be found in a number of genera of the two winged flies (Diptera) including the well known fruit fly *Drosophila*. Not all tissues in these insects possess such chromosomes, only certain very large cells in rectum, midgut, excretory organs and salivary glands. The structures are most frequently studied in the salivary glands of larval *Drosophila* and *Chironomus* (a group of non-biting midges). Slightly similar structures are found in some other animal and plant cells, for example the nucleus of the ciliate protozoan *Euplotes* during conjugation, but no other chromosomes are as amenable to light microscopical study as the Dipteran ones.

Polytene chromosomes are enormously large, but do not possess DNA molecules greatly longer than those found in many other chromosomes. They are large by dint of the fact that the DNA molecules remain fairly stretched linearly and are present as repeated copies lying side by side. In other words the DNA has replicated many times and the replicates remain in parallel linear array, with homologous pieces of chromosome in exact register. It has been estimated that the salivary gland polytene chromosome of *Chironomus* have arisen by 13 geometric doublings of the diploid set, and in most species they consist of hundreds or thousands of DNA molecules lying in extended parallel array. Actually each 'chromosome' consists of two homologous chromosomes lying side by side, and may often be seen to consist of two separate parallel bands, held together in most places but separate at others. The replicated copies of the same chromatid within one chromosome have been shown to exhibit an identical transcription pattern when present in an active gene complex (see McKnight & Miller 1979 and Fig. 6.6a).

Another point to note about these chromosomes is that, in many species, they are linked together at their centromeres, the centromeric junction forming a so called chromocentre. Two of the chromosomes in *Drosphila* are very long and have roughly central centromeres, namely chromosomes 2 and 3, so each of these appears in a squash with two long arms radiating from the chromocentre. Two others, the sex chromosome which is somewhat shorter, and the fourth chromosome which is very short, have centromeres almost at their ends (i.e. they are more or less telocentric) and so they appear as single arms. This implies that a really good squash preparation of these chromosomes will appear as in Fig. 6.7 with six arms, five of which are fairly long and all attached to a rough chromocentre.

A third and crucial point about these remarkable chromosomes, and the reason for their inclusion in this chapter, is that they persist throughout interphase and so are transcriptionally active. Indeed, it is relatively easy to dissect out the salivary glands from a *Drosophila* larva, incubate them in a tissue culture medium containing tritiated uridine, and permit the synthesis of radioactive RNA. If this is now followed by the preparation of autoradiographs from the squashed glands cells, evidence of widespread RNA synthesis on the chromosome bands will be obtained (see Fig. 6.8). The distribution of such RNA will be discussed in detail later.

6.2.1 Strandedness and Replication

As has been already discussed, polytene chromosomes are multistranded. But the situation is more complicated and also more interesting than that simple statement suggests. The truth is that not all parts of any one chromosome are involved in the same number of replications. This phenomenon was first noted in regard to the centromeric DNA which is, as in other organisms, simple sequence DNA and is strongly heterochromatic. In at least some *Drosophila* species the amount of such centromeric heterochromatin (see discussion in Chapter 7) reaches only the 4C level in the salivary gland cells—the amount expected in any normal G2 nucleus following S. By contrast, the rest of the chromatin is present at many times the 4C level (four times the haploid amount, i.e. the amount of DNA in a cell following the S phase of replication). So in truth the giant polytene chromosome does not in every respect consist of multiple copies of a normal chromosome in parallel array—only selected parts are highly replicated. The same theme applies to the genes for ribosomal RNA. These genes are clustered in the nucleolar organizing region of the X and Y sex chromosomes of *Drosophila*. In the normal diploid tissues of adult flies, these genes are present as very many copies and account for up to 0.5 % of the entire genome. Yet in the salivary gland tissue they account for only about 0.1 %. Their situation is however, less extreme than the centromeric heterochromatin,

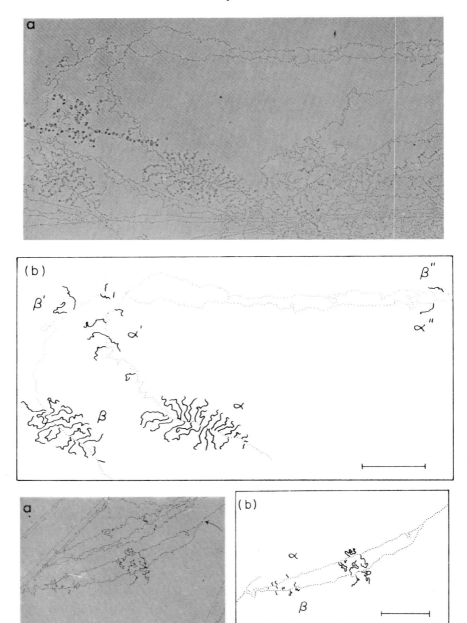

Fig. 6.6   (a) and (b) Electron micrograph of non-ribosomal transcription units on sister chromatids of *Drosophila* chromatin, together with interpretive drawing of micrograph in (b). (Kindly supplied by Dr McKnight & Prof. Miller.)

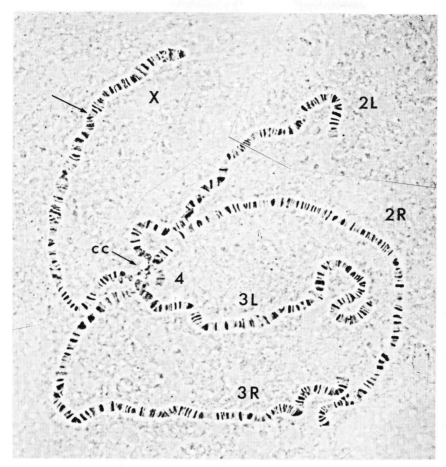

Fig. 6.7 Montage photograph showing the entire set of polytene chromosomes of *Drosophila*. Left and right portions of long chromosomes 2 and 3 are indicated, the X chromosome and the short chromosome IV. CC marks the chromocentre.

since they are present in greater numbers in the polytene chromosomes than in a 4C nucleus, but are under represented as compared with most of the genome. The numbers of ribosomal genes in the polytene chromosomes also seems to be under a very special kind of independent control. By genetic experiments with particular strains of *Drosophila*, it is possible to make strains with increased numbers of nucleoli and greater numbers of ribosomal genes in their genomes. But curiously, the salivary gland chromosomes of such flies have the normal number of ribosomal genes expected in the polytene chromosome. This surprising situation is part of the general phenomenon of 'gene magnification' or 'gene rectification' which applies to the ribosomal genes of Diptera. As we shall

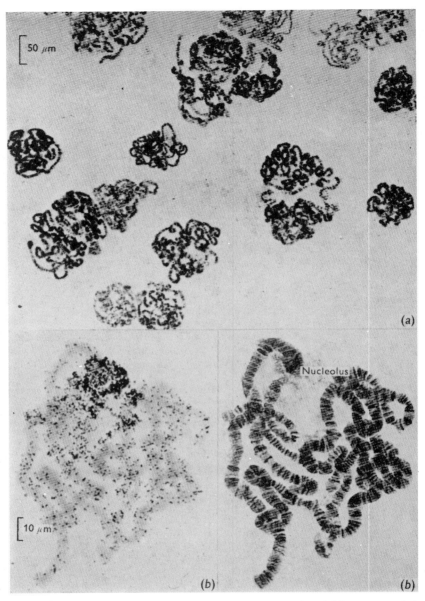

Fig. 6.8   (a) A squash preparation of a suspension of isolated nuclei from *Drosophila* salivary gland. (b) shows an autoradiograph and a phase contrast photograph of the same spread following a 20 min in-vitro incubation of the nucleus with tritiated UTP. Grains are indicative of UTP incorporation and RNA synthesis. (From Berendes H. D. (1971) *Symp. Soc. exp. Biol.* **25**, 145.)

see when we come to discuss chromosome bands in detail, it is also possible for individual bands to increase in length by an increase in their DNA, but this is not due to variable polytenization.

To summarize briefly, then, the giant polytene chromosomes consist of parallel replicates of the normal chromosomes but certain sequences are under-represented in the replication, some grossly so.

6.2.2   Bands, Interbands and Chromomeres

Even an unstained preparation of a squashed *Drosophila* salivary gland will demonstrate a banded pattern on the chromosomes, especially if viewed with a phase contrast microscope. Staining with acetic-orcein greatly improves the resolution of this banded appearance. The darker areas of the chromosome are referred to as the bands, the lighter portions as interbands. Bands vary considerably in length along the chromosome, some appearing as dense, thick bands and others as very thin indefinite lines. Moderate stretching of the chromosomes during squashing helps to resolve the banding pattern clearly and enables calculations to be made about the numbers of bands per chromosome and per nucleus. There are in fact about 1000 per chromosome arm and therefore about 5000 per nucleus.

The first important point to grasp about the banding pattern of these polytene chromosomes is that, in terms of both band number and size, it is constant in different tissues of the same fly, but varies for different species and genera of flies. In other words, it looks, even superficially, to be of genetic significance. More impressively, the two chromatids of any one chromosome may often be distinguished because the two sections of the chromosome become separated (known as asynapsis). In such a situation, as in Fig. 6.9(a), series of bands can sometimes be seen to be deleted or inverted in one chromatid and not in the other. Presumably such sections often fail to synapse. Such a situation may even be seen to apply to the form of one band in which the two chromatids lying more or less in register differ in the morphology of one band.

We can now proceed to ask how the DNA is distributed along the length of the chromosome. Is it evenly spread or irregularly condensed? The answer is an intriguing one. Although in terms of net length, the interband regions total almost half the chromosome length, it is found that only about 5% of the DNA is in the interbands and 95% in the bands (although some recent measurements of Laird put the interband proportion as high as 25%). It is therefore abundantly clear that the DNA must be highly condensed in the bands. Indeed, since the entire genomic length of *Drosophila* would provide about 5–6 cm of DNA (for the haploid genome), and the combined length of the polytene chromosomes is about 2000 μm, an overall packing ratio of 30:1 must be involved. A little arithmetic indicates that the interband DNA is scarcely condensed at all, while the band DNA is highly condensed, although still less so than in the DNA of a normal mitotic chromosome.

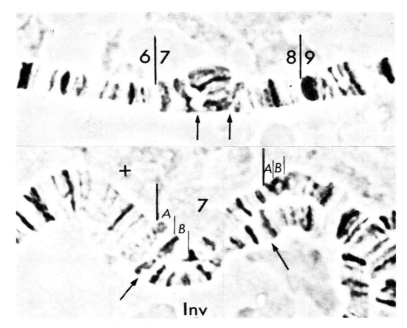

Fig. 6.9 A short inversion of the *Drosophila* X chromosome (Inv) with breakpoints at arrows. In the unsynapsed lower photograph the two chromatids of the set can be readily distinguished.

The condensed sections of DNA which form the bands are also known as chromomeres, a word which was applied also to the condensed DNA in the lampbrush chromosome. The interbands are therefore sometimes referred to as interchromomeres.

6.2.3 Where are the genes?

Since *Drosophila* has about 5 cm of DNA in its haploid genome, this amounts to about 500 million nucleotides. Since an average protein contains less than 1000 amino acids, and 3000 nucleotides code for 1000 amino acids, it follows that there is enough DNA in the *Drosophila* genome to code for at least 200 000 different proteins. Now 200 000 structural genes are almost certainly far more than *Drosophila* can need, or even sustain in terms of the possibility of mutation. So it comes as little surprise to find that, although there is space for some 30 or 40 structural genes in a polytene chromosome band, gene mapping studies suggest that there is most frequently only one gene per band. On this point the experimental evidence is clear, that the structural genes are located in the bands (although it is not possible as yet to exclude the possibility that the gene is actually at the band/interband junction), and that there is generally only one structural gene per band. This is not to say, of course, that an occasional band or chromosome might possess two separate structural genes and some might have none. But in general the rule is one gene, one chromomere or band. Of course, genes are here defined in terms of function, so that we are not excluding the possibility that some genes might be present within a chromomere as multiple copies.

A second line of evidence which underlines the close correspondence between chromomeres and genes as units of genetic function comes from studies on the puffing phenomenon and that will now be discussed.

6.2.4 Puffs on Polytene Chromosomes

It was stated earlier that the banding pattern of polytene chromosomes was essentially the same in different tissues of the same organism. That was something of an over generalization. For the truth is that some bands, although invariably present in all tissues, look dramatically different in certain tissues at certain times. This alteration in the appearance involves what has come to be known as puffing, an apparent decondensation of the chromomeric DNA to give a localized ring of diffuse material rather than a discrete band. Such objects were noticed by Balbiani in the 1880's and certain very large puffs on certain bands are still termed Balbiani rings. A puff seems to involve decondensation of all or most of the numerous DNA strands simultaneously within a band. Indeed some puffs are so large that a number of adjacent bands seem to be involved in the puff although it is difficult to determine whether one or more adjacent bands are truly puffed together. The fact that the same band can be seen to be puffed in one tissue and condensed in another is surely overpowering evidence that puffing is a visible sign of differential gene activity and indeed much of the visible 'halo' of a puffed region is not DNA but RNA and protein, the products of the localized transcription of the gene in question.

With the introduction of radioisotopes and autoradiography to biology, it was not long before research workers in a number of laboratories had shown that puffs were sites of intense RNA synthesis (see Fig. 6.10) and indeed that little RNA synthesis could be detected in parts of the chromosome that could not be designated as puffed chromomeres. (It is noteworthy that some bands which are active in transcription are not, however, noticably puffed.) So puffing seems to be visible evidence of gene activity and the giant polytene chromosome has presented science with the opportunity to view a gene in action underneath a light microscope! But let us look at the details of such transcription more closely. The first remarkable detail to underline is that often the entire chromomere is puffed and apparently transcribed, despite what has been said about the fact that the chromomere is extravagantly larger than a structural gene. It is still not clear how much of the decondensed chromomeric DNA is actually routinely transcribed into precursor messenger RNA. Some attempts have been made to isolate the RNA transcribed from particular polytene puffs. One such puff, termed Balbiani Ring 2 in *Chironomus*, and situated on the small chromosome IV, yields a very large RNA which sediments at about 75S (see Fig. 6.11). This is, of

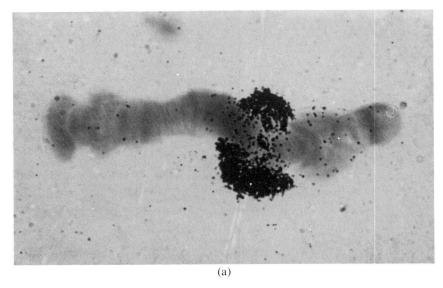

(a)

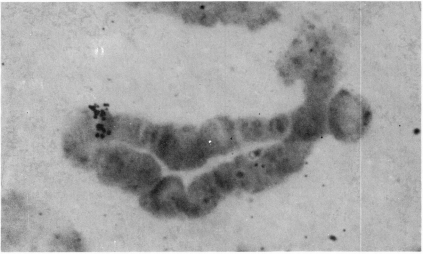

(b)

Fig. 6.10   The unfolded, transcriptionally active state (a) and the condensed, inactive state (b) of the BR 2 gene as demonstrated by in-situ hybridization of labelled BR 2 RNA to a salivary gland chromosome IV (a) and to a Malpighian tubule chromosome IV (b). (Photographs provided by Prof. B. Daneholt.)

course, much larger by a factor of 4 or 5 than any messenger RNA would be expected to be, and would be classed as a type of heterogeneous nuclear RNA. Even larger RNA, of a length equal to the entire chromomeric DNA, has been detected as a primary product of some puffs.

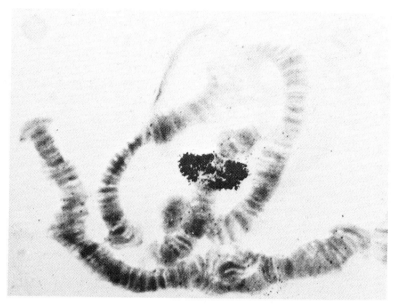

Fig. 6.11 Hybridization of radioactive labelled RNA with DNA in the Balbiani Ring 2 puff of *Chironomus tentans*. (Photograph kindly provided by Dr Lambert.)

Some other characteristics of polytene chromosome puffs deserve comment here. They are that the pattern of puffing on the chromosome set not only differs from tissue to tissue, but also, for any one tissue, varies from one developmental stage to another. Other factors such as temperature and hormone availability determine the activity or inactivity of particular chromomeres. This all adds up to the fact that puffing is at least a representation of differential gene activity. Genes which will be active in all tissues should be correlated with persistant puffs, while genes related to particular differentiated states and stages would be expected to have rather limited activities. Although really sound evidence on this topic is still rather scanty it is fair to say that such evidence as there is supports this view. What one can be more definite about is the total number of active chromomeres in a particular tissue. Ashburner has studied the puffing pattern of *Drosophila* larvae salivary glands during the third instar and the early prepupal period, and finds 108 loci forming puffs at one or both of these times. This may seem small as a measure of the total number of active structural genes, even if the overall number is little more than 5000. However, it is probable that some genes may be active but are scarcely if at all discernable as puffed chromomeres. Certainly autoradiography following incubation of salivary glands in tritiated uridine suggests RNA synthesis in some chromomeres which would not normally be designated as puffs in that tissue.

HEAT-SHOCK PUFFS

Many years ago it was reported by Ritossa that if *Drosophila* third instar larvae were subjected to elevated temperatures, the polytene nuclei in their salivary glands exhibited a specific set of puffs. Such puffs have come to be termed heat—shock puffs. The precise mechanisms underlying their appearance remains unclear but it is significant to note that cytoplasm from 'heat shocked' *Drosophila* tissue culture cells has been shown to induce the specific puffing pattern in isolated polytene nuclei (Compton & McCarthy 1978).

6.2.5   The Effects of Ecdysone

We have established that puffing represents gene activity, is more or less tissue specific, and can be induced or abolished in particular chromomeres by environmental factors. One such factor deserves closer scrutiny and further discussion, namely the insect hormone ecdysone.

Ecdysone is an important steroid hormone which stimulates moulting in insect larvae and helps to induce the morphological changes involved in metamorphosis to the pupa and adult. This hormone normally co-operates with another hormone, the so called juvenile hormone, to control the numerous changes which constitute insect metamorphosis. It is relatively easy to inject larvae between moults with purified ecdysone and to compare the salivary gland chromosome puffing pattern in these larvae with that of uninjected larvae. Alternatively, glands may be dissected out of larvae and incubated in experimental and control media with or without ecdysone. In the case of both *Drosophila hydei* and *Chironomus tentans*, numerous experiments in both English and German laboratories indicated that ecdysone would alter the puffing pattern, some previously inactive and unpuffed chromomeres becoming active (see Fig. 6.12) and other previously active ones becoming inactive. In short, ecdysone would visibly alter the particular programme of gene expression in the salivary gland. Moreover, the concentration of hormone seemed to affect the size of any individual puff and the rate of RNA synthesis within it.

The ecdysone experiments have been taken a little further and some of the proteins coded by the activated genes identified. A large mucopolysaccharide, used by the larva for salivary gland secretion, has been found to be specifically synthesized under the influence of the hormone, and there are good grounds for supposing that a number of the large Balbiani ring puffs on chromosome IV are responsible for its production.

Stimulation of visible gene activity by ecdysone sounds an ideal situation in which to determine the precise relationship between a hormone and a structural gene. It is tempting to believe that affected chromomeres have specific gene receptors, which bind ecdysone or an ecdysone/protein complex and so modify the expression of the rest of the chromomere. To date, the evidence is incomplete but strongly suggestive and we must await further experiments with labelled ecdysone and suitable salivary gland material. The response of different bands to the hormone is certainly variable, some responding quickly, others slowly, some to low concentration, others to high. Despite these limitations to our knowledge, however, the effect of ecdysone on the giant polytene chromosome, *in vitro* and *in vivo*, remains as the outstanding example of hormonal involvement in gene activity.

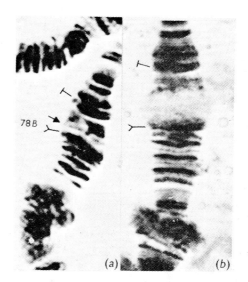

Fig. 6.12 Chromosomes of *Drosophila hydei* before (a) and after (b) exposure to ecdysone, showing induced puffing of the region designated at 78 B (from Berendes (1971) *Symp. Soc. exp. Biol.* **25**, 145).

6.2.6 DNA Sequencing and the Polytene Chromosome

A new perspective has opened up for the polytene chromosome with the advent of DNA sequencing studies. Such work is fairly recent and is fully discussed in Chapter 4 of this book. Only the results and insights relevant to these particular chromosomes will be included here.

Studies on the abundance and distribution of 'unique sequence' and 'middle repetetive' DNA in *Drosophila* have yielded somewhat conflicting results. There seems little doubt that, as in many other organisms studied, the 'middle repetetive' and 'unique sequence' DNA is interspersed in *Drosophila*. The main point of contention has been one of size. In many other organisms middle repetetive sequences of 0.2–0.5 kb (see discussion in Chapter 4) are freely interspersed among the unique sequences, the latter being 1–3 kb. In *Drosophila* most authors have found a rather different picture, using electron microscopy of reannealed DNA, following shearing and separation of unique and middle repetetive fractions. Authors report very long middle repetetive sequences, from 0.5 to 13 kb long, interspersed among very long unique sequences of some 13 kb and upwards. (The number of moderately repetitious sequences calculated would actually provide about one per chromomere if they were all completely interspersed with the unique sequences, and we will return to this point in a moment.) Long interspersed sequences are not unique to *Drosophila*, but have been detected in other, though not all, insects (e.g. *Apis*, the honey bee).

A second revelation of the research work on DNA sequencing in *Drosophila* is that palindromes are very numerous. A palindrome is an inverted repeat i.e. ABCABBACBA, and most frequently the mirror image repeats are on conplementary strands. Such sequences have a number of interesting properties. Not the least interesting is the potential ability of such sequences to form regions of localized double strandedness which could loop out from the main duplex and provide additional three dimensional complexity, perhaps as recognition sites by control molecules. Palindromic sequences seem to occur in all classes of *Drosophila* DNA, unique, middle repetetive and highly repetetive, but a greater proportion in the middle repetetive fraction. The number estimated for the *Drosophila* genome is 2000–4000, which, although probably not a very reliable figure, is again within the range of the total chromomere number.

We should now ask whether it is possible to incorporate this information about sequencing into a view of the polytene chromosome and especially its conspicuous subdivision into chromomeres. The highly repetitive DNA is known to be chiefly clustered at the chromocentre where the centromeres are aggregated. The so called moderately repetitious sequences are presumably not clustered and indeed some middle repetitive families are widely dispersed, but their repetitious nature means that something like each chromomere could have a section of its DNA, perhaps a large section, very similar to such sequences in other chromomeres. Each chromomere may also possess a palindromic sequence and this could, at least in most cases, be the same middle repetive sequence shared by other chromomeres. Such frequently occurring palindromes could easily function as important control regions.

6.2.7   Variation in DNA Content

The point has already been made in section 6.2.1 that polytene chromosomes can vary in their overall DNA content as a result of an under-replication of some of the genes, particularly the genes coding for ribosomal RNA. But another curious phonomenon has come to light through studies on these chromosomes and this involves variation in total length of such chromosomes, rather than variation in breadth (i.e. in degree of replication). Such variation is not randomly distributed but is a particular feature of individual chromomeres, and is detectable when closely related sub-species of Diptera are compared. The best example involves work done by Dr H. G. Keyl on two species of *Chironomus*, in which he found that corresponding bands in the chromosomes differed in their relative DNA content. Moreover, the differences were in definite fixed ratios of 2, 4 or 8 fold. This evidence seems to demonstrate that individual chromomeres can be replicated separately and that copies of the entire chromomeric DNA may be integrated by accident or intention. It is hard to believe that such double or quadruple bands have much to do with the essential genetic endowment of the fly.

6.3 MODELS OF CHROMATIN ORGANIZATION. THE LAMPBRUSH VERSUS THE POLYTENE

As was underlined at the beginning of this chapter, normal chromatin cannot be visualized in the light microscope during transcription. This difficulty largely explains the present state of ignorance about the organization and functional activity of chromatin. It follows that the lampbrush and polytene chromosomes have attracted great attention not only out of interest in their particular structure and function but because they can be viewed as models, or at least indicators, of how eukaryotic chromatin actually functions. Much of this will be discussed in Chapter 7 of this book, but here we will attempt a comparison of the lampbrush and polytene chromosomes in preparation for that discussion. Essentially both structures consist of a string of chromomeres, of which the overall number is a few thousand, averaging say 5000 for some Amphibian lampbrush and some Dipteran polytenes. In both, each chromomere consists of highly condensed DNA and some 95% of the total DNA is contained in the chromomeres. The linker DNA is short, uncondensed and possibly untranscribed in each type. But there is a marked dissimilarity between the two structures in the mechanism of chromomeric transcription. In the lampbrush a short loop is extended, long enough for only one or a very few genes, while most of the basal chromomeres remain condensed. Transcription seem to involve only the loop, or even a part of a loop. Polytene transcription, however, appears often to involve decondensation and reading of most or all of the entire chromomere, a section of DNA long enough to code for some 30 to 40 genes. A second important difference is closely implicated with this first one. It is that in the lampbrush all, or almost all, chromomeres are actively transcribing since all seem to have loops, while in the polytene only a few bands are puffed, implying transcription from only a maximum of a few hundred chromomeres, perhaps far fewer. There, then, are the bare bones of the dilemma. In the one a small part of each chromomere is being simultaneously transcribed, in the other the whole of a very small number are active together, while most chromomeres are entirely condensed and inactive. Which is the correct model of eukaryotic transcriptional control?

Now let us be suitably cautious. On the one hand, some of the generalizations we have arrived at could be misleading, especially the inactivity of the linker DNA joining the chromomeres and even the vexed question about whether the loop DNA in the lampbrush moves, being spun out and gathered in continuously. But given that the generalizations are broadly true and that at least one of these structures is broadly typical of what goes on, we are left with a tantalizing choice.

Could it be that in interphase chromatin, most or all structural genes are actually being constantly transcribed, throwing the weight of control of gene expression on post-transcriptional selection of appropriate messages? That is what the lampbrush seems to suggest. It could also be taken to suggest that most of the genome is rarely if ever transcribed. That would imply partial transcription at any one time, perhaps total transcription over a long period and still leave the onus of regulation to be handled post-transcriptionally. But it is important to stress that the lampbrush chromosome is meiotic and its transcription pattern may be purely characteristic of that stage. Polytene chromosomes in germ line cells are reported to have an incorporation pattern reminiscent of the lampbrush.

Taking the polytene as a model yields a more conventional view, perhaps, but still leaves problems. In this conception only a few genes are transcribed at any one time, making transcriptional regulation the key to selective gene expression. But it also implies that a whole complex of sequences within a chromomere is simultaneously decondensed and probably, though not necessarily, transcribed. Whether the intron/exon organization of genes will clarify this question is not yet clear.

Both chromosomes tend to suggest that the chromomere is a unit of transcription and that there is roughly a one structural gene–one chromomere relationship. Or at least that is the direction in which most of the present evidence points. We will now defer to the next chapter the wider issue of a general discussion on eukaryotic gene expression and its regulation.

6.4   FURTHER READING

Ashburner M.& Richards G. (1976) The role of ecdysone in the control of gene activity in the polytene chromosomes of *Drosophila*. In Lawrence P. A. (ed.) *Insect Development.* Blackwell Scientific, Oxford.

Beermann W. (ed.) (1972) *Developmental studies on giant chromosomes.* Springer-Verlag, Berlin.

Bridges C. B. (1935) Salivary chromosome maps *J. Hered.* **26**, 60–4.

Callan H. G. (1967) The organization of genetic units in chromosomes. *J. Cell Sci.* **2**, 1–8.

Callan H. G., Gross K. W. & Old R. W. (1977) Localization of histone gene transcripts in newt lampbrush chromosomes by *in situ* hybridization. *J. Cell Sci.* **27**, 57–80.

Compton J. L. & McCarthy B. J. (1978) Induction of the *Drosophila* heat-shock response in isolated polytene nuclei *Cell* **14**, 191–201.

Gall J. G. & Callan H. G. (1962) H³ uridine incorporation in lampbrush chromosomes. *Proc. Natl. Acad. Sci. U.S.A.* **40**, 562.

Harford A. G. (1977) The organization of DNA sequences in polytene chromosomes of *Drosophila*. In Li & Eckhardt (eds) *Chromatin and Chromosome Structure.* Academic Press, London.

Keyl H. G. (1965) A demonstrable local and geometric increase in the chromosomal DNA of *Chironomus*. *Experientia* **21**, 191–9.

Lefevre G. Jr (1976) A photographic representation and interpretation of the polytene chromosomes of *Drosophila melanogaster* salivary glands. In Ashburner M. & Novitski E. (eds) *The Genetics and Biology of Drosophila*. Academic Press, New York.

Macgregor H. C. (1977) Lampbrush chromosomes. In Li & Eckhardt (eds) *Chromatin & Chromosome Structure*.Academic Press, London.

Macgregor H. C. (1980) Recent developments in the study of lampbrush chromosomes. *Heredity* **44**, 3–35.

McKnight S. L. & Miller O. L. Jr. (1979) Post-replicative nonribosomal transcription units in *D. melanogaster* embryos. *Cell* **17**, 551–63.

Mott M. R., Barnett E. J. & Hill R. J. (1980) Ultrastructure of polytene chromosomes of *Drosophilia* isolated by microdissection. *J. Cell. Sci.* **45**, 15–30.

Mott M. R. & Callan H. G. (1975) An electron microscope study of the lampbrush chromosomes of the newt *Tristrus cristatus*. *J. Cell Sci.* **7**, 241.

Sheer U., Franke W. W., Trendelenburg M. F. & Spring H. (1976) Classification of loops of lampbrush chromosomes according to the arrangement of transcriptional complexes. *J. Cell Sci.* **22**, 503–19.

Cold Spring Harbor Symposium on Quantitative Biology. (1977) *Vol. XLII, part 2*. (Numerous Papers.)

Chapter 7
Chromatin Activity and Gene Regulation

In the preceding chapters of this book the subject has been discussed mainly from the point of view of structure. In this chapter the emphasis is on the problem of how chromatin functions. It can be briefly stated that chromatin is active in two quite distinct ways, in self replication of its DNA on the one hand, and synthesis by chromatin, the phenomenon of transcription. RNA synthesis is been outlined in Chapter 5, leaving us now to deal exclusively with RNA synthesis by chromatin, the phenomenon of transcription. RNA synthesis is termed transcription because it involves a copying of the DNA code, with the minor adjustment of uracil replacing thymine and a change of sugar in the molecule, so that the RNA produced is precisely complementary in its base sequence to the DNA strand copied. The later step of translation consists of a radical shift from the language of nucleotide bases into the language of amino acids on a 3 to 1 basis. Clearly translation of RNA and the production of protein is itself an important aspect of gene regulation. Indeed it is the level at which the topic has been most diligently studied. But since the whole subject of protein synthesis is both broad and complex, in this book gene expression will be discussed chiefly in terms of transcription and RNA synthesis, a few paragraphs on the topic of post-transcriptional control being added at the end of the chapter.

7.1 THE MECHANISM OF TRANSCRIPTION

RNA is synthesized by the activity of a group of enzymes, the RNA polymerases. These enzymes are much less well defined structurally in eukaryotes than is the RNA polymerase of bacteria, but in both prokaryotes and eukaryotes each enzyme is made up of a number of subunits with a combined molecular weight of about 500 000 daltons. So RNA polymerase is quite a hefty enzyme. At least four main types of RNA polymerase occur in eukaryotes, each being responsible for catalysing the production of a different class of RNA (see Table 7.1). Enzyme A is chiefly associated with the nucleolus and is apparently solely responsible for making ribosomal RNA. Type B enzyme, which can be recovered from nucleoplasm, synthesizes the non-ribosomal species of RNA, with the exception of 4S (transfer) and 5S (Small ribosomal) RNA, which are synthesized by the third enzyme, type C. There is some indication that the type C fraction actually includes two fractions, so perhaps the 4 and 5S RNAs each have a separate polymerase. A fourth distinct enzyme is responsible for the synthesis of RNA within mitochondria.

Table 7.1   General Properties of Eukaryotic DNA-Dependent RNA Polymerases.
(Data from Dr P. H. W. Butterworth and others.)

Type	I(A)	II(B)	III(C)
Location	Nucleolus	Nucleoplasm	Nucleoplasm + cytoplasm
α-amanitin sensitivity	insensitive	10^{-8}–10^{-9}M	10^{-4}–10^{-5}M
M^{2+} preference	Mg = Mn	Mn	Mn $\geqslant$ Mg
$(NH_4)_2SO_4$ optima	0.04M	0.06–0.13M	0.05/0.16M
DNA template preference	ds	ss	ds = ss
Function (?)	rRNA	mRNA	4 + 5S RNA

A fortuitous discovery has greatly benefited research on RNA polymerase enzyme, namely that a notoriously poisonous mushroom, the death cap, *Amanita phalloides*, owes its toxicity to an octapeptide molecule, α-amanatin, which specifically inhibits the activity of the B type RNA polymerase of eukaryotic cells (it does, at much higher concentration, have some inhibitory effects on the type C enzyme also). Use of α-amanatin permits selective experiments to be carried out in which only synthesis of ribosomal RNA is permitted to occur. Another drug, Actinomycin D, complements the use of α-amanatin, in that it inhibits ribosomal RNA synthesis fairly specifically.

When a gene is to be transcribed, the first step is the attachment of the RNA polymerase enzyme to the appropriate region of DNA (see Fig. 7.1). This is accomplished by binding of the polymerase enzyme to a region of the particular complementation group (that is, the whole length of DNA responsible for production of the specific polypeptide chain) known as the promoter region. Binding of the enzyme to the promoter is affected by many factors, including cyclic AMP, and often the form of the polymerase molecule itself. This last comment refers to the fact that, at least in bacteria, a subunit of the entire enzyme, known as the sigma factor, alters the affinity of the enzyme for particular promoters. Having once attached to a promotor, the enzyme can proceed to engineer the assembly of ribonucleotides in a sequence complementary to the DNA base sequence of the gene. In doing so, the polymerase enzyme traverses the gene, moving from the start or initiation site to the end or termination site. In recent years electron microscopy has enabled us to visualize transcription in process, both of ribosomal and non ribosomal genes, and some of these can be seen in Fig. 7.2, 7.3 and 7.4. We can see in these remarkable photographs the 'Xmax trees' of RNA (referred to in Chapter 6) resulting from the increasing length of the RNA chain as the enzyme traverses the gene. Indeed, in some favourable pictures, the polymerase molecules themselves can be visualized on the DNA thread. The correct sequence assembly of the ribonucleotides is thus achieved by the RNA polymerase activity.

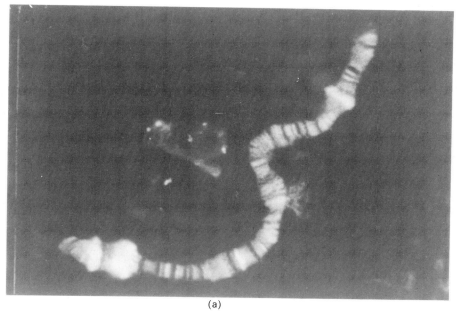

(a)

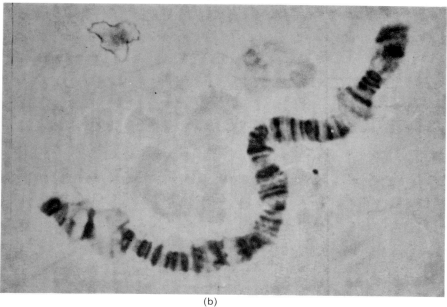

(b)

Fig. 7.1 Immunofluorescent labelling of *Drosophila melanogaster* RNA polymerase in a fragment of salivary gland chromosome from a heat shocked larva. (a) is with label, (b) is the same section viewed in phase contrast after orcein staining. (Photographs kindly provided by A. Greenleaf, V. Plagens and E. Bautz and reproduced by permission of Springer-Verlag, Berlin and editors of *Chromosoma*.)

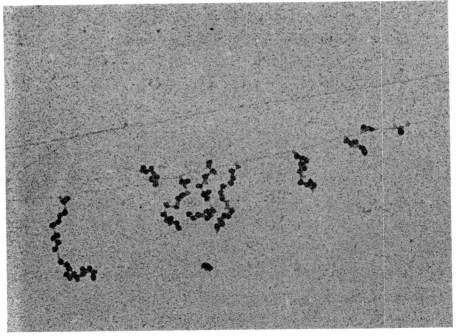

Fig. 7.2 Electron micrograph of the transcription and translation processes in bacteria. Two stretches of DNA, one naked, the other with nascent messenger RNA arranged at right angles, are observed. The bottom diagram facilitates the interpretation of the electron micrograph. (Courtesy of Prof. O. L. Miller.)

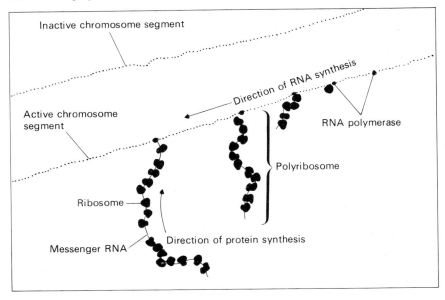

7.2 THE SELECTIVITY OF TRANSCRIPTION

Electron microscopy has shown (Figs 7.3(b) and 7.4) that not all the DNA is being transcribed all of the time, and indeed it is fairly certain that only small portions of the total genome are being copied in any particular cell at any one point in time. The details of just how gene transcription comes to be so selective will now be considered.

7.3 UNTRANSCRIBED DNA

Part of the DNA in chromatin seems to consist of sequences which are never transcribed during the normal life of the cell. Into this category of untranscribed DNA comes the centromeric heterochromatin yielding satellite DNA in the ultracentrifuge, spacer regions in the nucleolar genes, and probably the spacer regions which are the interbands of polytene chromosomes (see Chapter 6). Unfortunately, it is exceedingly difficult to be certain about which tracts of DNA are never transcribed *in vivo*, partly because of the present inadequacy of available experimental methods, and especially because transcription *in vitro* cannot be taken to imply transcription *in vivo*. It is clear that such transcription may not mirror exactly the pattern of gene copying in the normal eukaryotic cell.

What is implied in the paragraph above is that for some of the DNA, neither strand of the double helix is ever transcribed. What is also certain is that, for most eukaryotic DNA, only one strand is read, but which strand varies as one progresses along the molecule. The strand which is read is termed the sense or message strand, its complement the nonsense or antimessage strand. Since the complement of the promoter sequence does not itself promote, the likelihood of both strands being read from the same initiation point is negligible. But this does not rule out the possibility of some overlapping transcription of complementary strands, and this is indeed known to occur in many phage. Reading of complementary strands will of course proceed in opposite directions and will yield totally distinct messages, albeit with some tracts of complementary sequences (see also discussion on p. 210 and Fig. 6.4).

7.4 SELECTIVE READING OF TRANSCRIBABLE DNA

A large proportion of the DNA of an organism is, beyond doubt, potentially transcribable. Yet it has been recognized for many years that all cells and organisms exercise strict control over which genes are read and at what time. When the phenomenon of cell differentiation was first appreciated it might have been imagined that this cell specialization was achieved by the total discard of those parts of the genome not relevant to that cell, and the retention of only

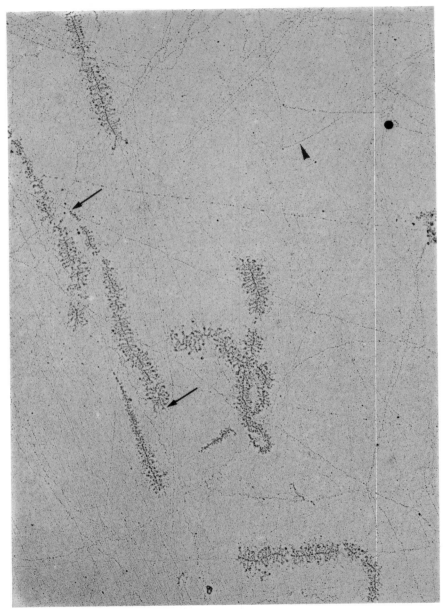

Fig. 7.3 (a) Transcriptional complexes, 'Xmas trees,' interpreted as nucleolar type visible within a network of inactive DNP fibers. The arrows (→) indicate the limits of a transcriptional unit which are of variable length and do not show regular spacers. Beaded aspect of DNP fibres (>) is probably due to the shadowcasting with platinum ('decoration') × 16 000.

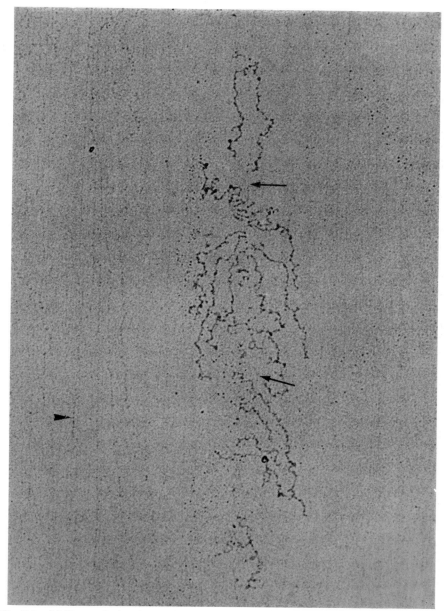

Fig. 7.3 (b) Transcriptional complexes of the non-nucleolar type. Very irregular and loose spacing of the highly twisted, lateral fibers along the faintly stained DNP axes (→). The longest of these RNP branches measures 1.7 μm. Many other 'silent' DNP fibers with granular appearance (>). Positive stain with PTA. ×34400. (Photographs kindly supplied by Prof. F. Puvion-Dutilleul. (From Puvion-Dutilleul F. *et al.* (1977) *J. Ultra. Res.* **58**, 108).)

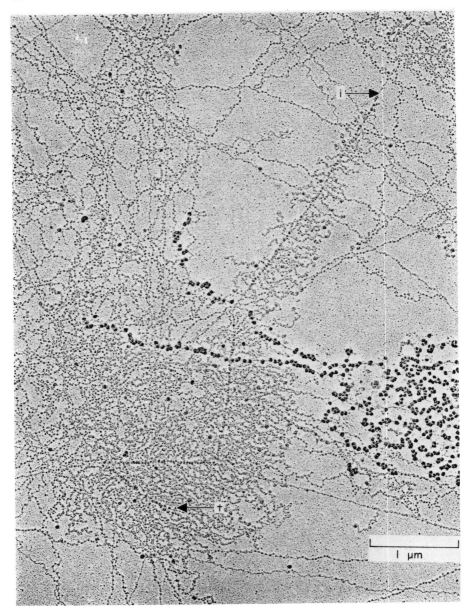

Fig. 7.4 Electron micrograph of a putative silk fibroin transcription unit. The Arrows point to sites of initiation (i) and termination (t) of transcription. Contour length (i) → (measures 5.3 μm.) The template is complexed with—10 RNA polymerase molecule per micrometer of contour length. (Photograph kindly supplied by Prof. D. Milier and Dr S. Knight.)

those genes relevant to that particular cell's destiny. All such genes would then be freely active. But nature has adopted the opposite strategy. Almost all cells in all organisms are endowed with, and retain, two sets of the full complement of genes which comprise the genome of that organism (the exceptions include the haploid cells of sperm and ovum, and certain examples of chromosome elimination found in gall midges, *Cecidomyidae*, the protozoans *Stylonichia* and *Oxytricha*, and the nematode worm *Parascaris*), leaving the onus of responsibility for restricted gene expression on the capacity of each cell to control the activity of its own DNA. Even in the simplest organisms such as bacteria, and the subcellular factors, the viruses, gene expression is tightly controlled so that while some genes are active at one time, others are silent. To appreciate the extent of the control involved, one has only to refer back to the giant polytene chromosomes discussed in Chapter 6. As was stressed there, certainly not more than a tenth of the potentially active genome is apparently transcribed at one time (see Fig. 7.3(b)).

Since selective gene expression is the way in which cells utilize their DNA, we will not be surprised to find that such selectivity operates in a number of different ways and at different levels.

7.5 POSITIVE AND NEGATIVE CONTROL

It is useful to stress at the outset that control could operate in two ways. If a gene, complete with its promoter, is freely available to the polymerase enzymes of the cell but can only be transcribed by the additional co-operation of some special molecule or factor, this is termed positive control. The special regulatory factor then positively turns on the activity of that gene. If, on the other hand, a gene exposed for transcription is automatically transcribed, the cell must take good care to prevent its transcription by denying access to the polymerase molecules or ensuring their immobilization on the sequence. Such regulation is termed *negative* control, implying that, for a gene to be read, some control factor must be removed or derepressed. Most known examples of gene regulation are examples of negative control. We can now examine the different tactics employed by cells to achieve selective gene expression, and will do so by thinking first of all at the gross level and coming down gradually to the selective activity of an individual gene.

7.6 CHROMOSOME INACTIVATION

7.6.1 Heterochromatin and Euchromatin
Although, as was stressed earlier, control of gene expression by actual elimination of genes is rare, there are a few examples of a phenomenon only slightly less

final in silencing gene activity. This is the permanent inactivation, in a cell, or a line of cells, of whole chromosomes, sometimes one, sometimes an entire set. Such inactivation is achieved by the permanent condensation of such chromosomes, which are then said to be heterochromatic. An elaborate terminology has arisen in connection with the different functional and structural states of chromatin, and these must be briefly discussed. When the terms euchromatin and heterochromatin were first employed they were most commonly applied to entire chromosomes, the sex chromosomes being referred to as heterochromosomes because one X chromosome in the female and the Y chromosome in the male were permanently condensed. But more recently heterochromatin has been defined by reference to three criteria—permanent condensation throughout interphase, genetic inactivity, and DNA replication late in the S phase, as compared to other chromosomes. This topic is discussed at length by Comings in his 1972 review. By contrast, euchromatin is chromosomal material that does not have these characteristics. Perhaps unfortunately, some investigators and writers have seen fit to use these terms to apply to material which satisfies only one or two of these criteria. Even the areas of condensed chromatin visible in electron micrographs of most eukaryotic interphase nuclei are sometimes referred to as heterochromatin. This is unfortunate but seems to have become accepted usage. More detailed subdivision of the term heterochromatin has, however, proved necessary and the most commonly adopted are the epithets constitutive and facultative. The former term is used to refer to chromatin which is permanently condensed in two homologous chromosomes—for example the centromeric heterochromatin, while facultative heterochromatin occurs on only one of two homologues, for example the inactivated X in the human female. Even this classification does not solve the issue entirely since the human Y chromosome is permanently condensed but has no homologue. Some writers refer to chromosomes such as the human Y as examples of semifacultative heterochromatin.

Probably it is best to use this terminology in the simplest possible way, in which constitutive heterochromatin is condensed chromatin which is never used as a source of genetic information, e.g. centromeric heterochromatin, and facultative heterochromatin is temporarily unused chromatin, but is utilized in some cells at certain times, e.g. condensed sex chromosomes. Euchromatin is genetically and transcriptionally active chromatin although much of it is only active for short periods of time. It includes chromatin that is often condensed in certain situations—most chromatin of nucleated erythrocytes, of polytene and lampbrush chromomeres, and also the entire conglomerate of RNA genes, control genes, spacer sequences and structural genes which make up the bulk of most chromosomes and chromatin.

Having outlined the slightly complex nomenclature, let us resume discussion of examples of inactivated chromosomes. The best known example is the inactivation of one copy of the X chromosome in the normal human female. Its condensation during interphase permits its visualization as the Barr body, which provides the basis for rapid screening of the chromosomal sex of individuals. The system of inactivating one X chromosome is found in all female mammals, although in marsupial mammals the X chromosome derived from the father is exclusively inactivated. Two points about X chromosome inactivation are well established and of considerable interest. The first is that, where inactivation does occur, it takes place very early in embryonic life, although both X chromosomes are clearly active at the very earliest stages of female development. This fact is demonstrated by the abnormality of Turner's syndrome (a syndrome present in human XO females, who display short stature and some anatomical aberrations) in humans, which is an XO genotype. If only one X chromosome were active at all stages of normal human female development, an XO genotype should lead to normality. The aberration of Turner's syndrome obviously follows from lack of the genetic activity of two X chromosomes very early in embryonic life.

The second point is that, except in marsupials, the paternal and maternal X are inactivated randomly in early embryonic life. In later life, however, clones of cells appear which all have the same X chromosome condensed. This implies that animals heterozygous for genes carried on the X chromosome will be natural mosaics in the female. Such indeed is the case with the coat colour genes of the tortoise-shell cat, which is, of course, invariably female.

X-chromosome inactivation in female mammals has been cited by Dr Mary Lyon as an example of gene dosage compensation (often referred to as the Lyon hypothesis), on the grounds that, if numerous X chromosomes are present in the genotype, all but one are inactivated. As shown in Fig. 7.5, if an individual is XXX, XXXX, or XXXY, the number of Barr bodies visible is always one less than the total number of X chromosomes. Curiously, in some triploids and tetraploids, more than one X chromosome is normally active, suggesting that the dosage is related in some way to the overall chromosome complement.

As well as one of the female X chromosomes being heterochromatic in most mammals, the male Y chromosome is also. As is stated above, the precise nature of Y chromosome material is not easy to determine, since it is highly condensed yet, in some ways, genetically active. After all, it is male determining, since human males are very different from Turner's syndrome XOs who are phenotypically female. Actually in insects and many other organisms, the Y chromosome is not male determining and XO individuals are male and XXY individuals female.

So far, only chromosome inactivation involving the so called sex chromosomes has been discussed, but it is not confined to these structures. In the mealybugs, Coccoidea, female insects have ten functional and uncondensed chromosomes, while the males have five of the ten chromosomes permanently condensed. Not only does this mean that the males are functionally haploid, but curiously enough the five condensed chromosomes are invariably those which have been

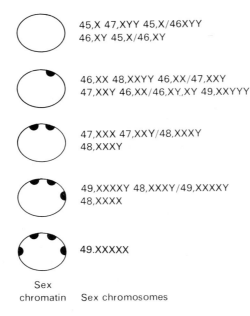

45,X 47,XYY 45,X/46XYY
46,XY 45,X/46,XY

46,XX 48,XXYY 46,XX/47,XXY
47,XXY 46,XX/46,XY,XY 49,XXYYY

47,XXX 47,XXY/48,XXXY
48,XXXY

49,XXXXY 48,XXXY/49,XXXXY
48,XXXX

49.XXXXX

Sex
chromatin   Sex chromosomes

Fig. 7.5 Schematic diagram to show the correlation between the number of X chromatin masses, (Barr bodies) and number of X chromosomes. (After Hamerton J. L. (1971) *Human Cytogenetics*. Academic Press, New York.)

paternally derived. Another example of condensation of autosomes is found in certain extra so called supernumerary of B chromosomes, encountered in maize, rye, grasshoppers and many other animal and plant species. These B chromosomes are still poorly understood since, in most cases, they appear to be entirely dispensable genetically; genetic inactivity and severe condensation are both common features of such chromosomes.

7.6.2   Mitotic Chromosomes

There is, of course, one time in each cell cycle when all chromosomes are severely condensed (but, as has been stressed, they should not be referred to necessarily as heterochromatic), and that is during mitosis. This is also a time when gene activity in terms of RNA synthesis is silenced. Although chromatin condensation at mitosis is no doubt a result of the requirement for separate packaging to ensure an exact distribution of chromatin between the daughter cells (see discussion of mitosis in Chapter 5), it is important to notice that this condensation is apparently incompatible with transcription and RNA synthesis is suspended. Although a continuation of RNA synthesis during mitosis might well be good for the cellular economy, the severe condensation apparently prevents it. It is plausible to conclude that the chromosome condensation which affects the inactivated X chromosome in Barr body formation is analogous to mitotic condensation, but is more permanent.

7.7 REGULATION AT A GROSS LEVEL

7.7.1 Control of Large Tracts of Chromatin

The only common example of a large block of chromatin of less than chromosomal size which is selectively inactivated is that of the constitutive heterochromatin already referred to. This material consists of simple sequence DNA, is commonly present as a block of centromeric chromatin, but is also present in small blocks around the nucleolar organizing regions, the 5S RNA regions and at the ends of certain chromosomes. Such chromatin in non-centromeric sites is referred to as intercalary heterochromatin and that at the tips of chromosomes as telomeric heterochromatin. No examples are known of facultative heterochromatin existing in blocks smaller than entire chromosomes, and so it must be clear that these examples do not truly represent a deliberate switching off of gene activity by the cell but rather of an insertion within chromatin of lengths of DNA which cannot normally be transcribed. There is one other interesting comment to make here, however, and it is that certain genes, normally active in a particular type of cell, are found to be inactivated if they have been translocated to a new position adjoining a piece of heterochromatin. One of the earliest recognized examples of this phenomenon was the phenotypic appearance of the *Drosophila* eye, in which some facets may be white rather than red. The eyes of affected flies are normally variegated, some facets being red and others white, apparently resulting from inactivation of the eye pigment gene which produces the red eye. Variegation seems to result from variation in the inactivating influence of the heterochromatin in different cells. This phenomenon is known as variegated position effect, and apparently does not necessarily demand direct proximity of the affected genes to the heterochromatin (for further discussion of position effect see Lewis 1965). The phenomenon also occurs when genes are translocated onto an inactivated X chromosome.

7.7.2 Controlled Expression of Homologous Tracts

Although still poorly understood and not necessarily even transcriptional in its basis, it is appropriate to comment here about a remarkable finding made by Honjo and Reeder (1973) in crosses of *Xenopus laevis* with the closely related species *Xenopus mulleri*.* These species of *Xenopus* have distinct types of ribosomal RNA, but in the hybrid, no matter which species is the male or the female, only *X. laevis* type ribosomal RNA can be detected. These ribosomal genes do of course exist in long tracts, and the expression of one tract is, in some way, prevented by the presence of the *X. laevis* genes on the homologous chromosomes. A somewhat similar phenomenon involving the expression of single immunoglobulin genes occurs in mammalian antibody forming cells and is known as allelic exclusion. It will be discussed later in this chapter.

*This species now styled *Xenopus borealis*.

7.7.3   Chromatin Condensation in Interphase Nuclei

Some interphase nuclei possess chromatin which is very severely condensed, and no better examples of such dramatic condensation exist than the nuclei of bird or amphibian erythrocytes (see Fig. 7.6). As we might expect, the levels of RNA synthesis in these nuclei are also very low. Here then are examples of a phenonemon which no doubt applies to a lesser degree to many nuclei—the shutting down of gene activity as a result of condensation of the chromatin. Although, when discussing mitotic chromosomes, it was stated that their condensation was presumably not engineered in order to accomplish the silencing of the genes, in the cases now being discussed, this may well be the case. Admittedly the chromatin packing in the chicken erythrocyte involves a novel species of histone, histone 5, and this histone is not recorded for other nuclei which are less severely condensed. But, whatever may be the mechanism of condensing interphase chromatin—and we are largely ignorant of how it is achieved—in at least some cases it seems to be used simply to bring about the inactivity of certain genes present in the condensed material.

7.8   CONTROL OF INDIVIDUAL GENE SEQUENCES

Since the genomes of prokaryotes are much smaller and simpler than those of eukaryotic organisms, it is hardly surprising that our present knowledge of transcriptional control is very much greater for bacteria than for animals and plants. Moreover, in the face of considerable ignorance of eukaryotic gene control, ideas on the subject have been considerably influenced by knowledge of the process in prokaryotes. Whether such an extrapolation is justifiable may seem very questionable, but we can best begin discussion of control of the expression of individual sequences by recalling what is known to occur in bacteria, as described more fully in Chapter 3.

As far as is known, bacteria do not resort to any of the devices already alluded to in this chapter as being relevant to eukaryotic gene expression, i.e. permanent or impermanent heterochromatin, or the severe condensation of chromatin in such cells as chicken erythrocytes. Instead, bacteria have developed a system of co-ordinate gene expression by which groups of genes coding for proteins with interrelated functions (assuming that interrelation of function requires that such proteins should all be available in the cell at the same time) are all controlled as a transcriptional unit. The existence of such functional units of transcription was first suggested by the French workers Jacob and Monod, based on their brilliant experiments on induced enzyme synthesis. Such a unit of transcription they termed an operon.

7.8.1   Bacterial Operons

See discussion in Chapter 3.

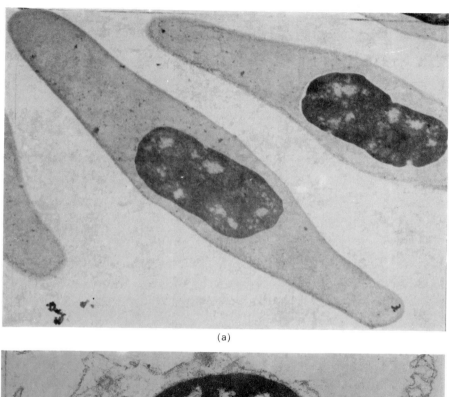

(a)

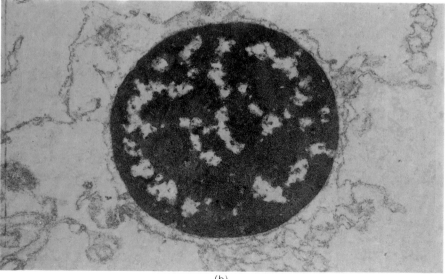

(b)

Fig. 7.6 (a) Electron micrograph of *Xenopus* erythrocytes. Notice the highly condensed chromatin in the nucleus. (b) Electron micrograph of isolated nucleus from *Xenopus* erythrocyte, again showing highly condensed chromatin. (Both photographs kindly provided by Dr H. Chegini.)

7.8.2 Sequential Control of Gene Expression in Bacteria

A mechanism which has often been suggested as a control mechanism in gene expression is one based on gene order and dependence of one sequence on the product of another 'earlier' sequence. An excellent example has been found in the transcription and translation of the small bacteriophage T7 within *E. coli*. When this small DNA phage enters the bacterium, only some of its genes can be transcribed by the *E. coli* RNA polymerase. These 'early' genes include one which codes for a novel viral RNA polymerase. This new polymerase then has the facility to transcribe certain 'late' viral genes which are refractory to the bacterial polymerase. No reports of similar mechanism operating on the bacterial genome have appeared, but it is an interesting mechanism to bear in mind when considering the eukaryotic situation.

7.8.3 Multiple gene Copies in Bacteria

Since the entire genome of *E. coli* is sufficient to code for only about 4000 different proteins, it is obvious that there is little scope for bacteria multiplying up the number of copies of an individual gene sequence in order to boost the output of its product. Only in the case of genes coding for ribosomal and transfer RNA is there any evidence for repeated gene sequences in prokaryotes. Indeed *Mycoplasma*, smaller and simpler than a bacterium, possesses only single copies of the genes coding for its 16s and 22s ribosomal RNA. But *E. coli* carries six cistrons for 16s and 23s ribosomal RNA and a larger number for the smaller 4s and 5s RNA's. As in eukaryotes, the primary product of the ribosomal gene sequences is a polycistronic RNA which is later cleaved and trimmed to yield the functional ribosomal RNA.

7.8.4 Control of Individual Sequences in Eukaryotes

Despite the great interest in the question of how genes are turned on and off in eukaryotes, our knowledge of this topic is still at a very preliminary stage. Perhaps the first point to be stressed is that there is no good evidence for true operons functioning in eukaryotes. The examples of the co-ordinate production of different proteins which have been discovered in eukaryotes are all affected by two qualifications. The first is that they involve increases in production of certain proteins but not a complete de-novo synthesis. This compares dramatically with induced enzyme synthesis in bacteria. Secondly, the known examples are all most easily explained as co-ordinated responses of a number of separate genes to a single effector such as a hormone. To take an example (see Table 7.2), the hormone glucagon initiates increased production of phosphoenolpyruvate carboxylase, tyrosine aminotransferase, L-serine dehydratase and glucose 6-phosphatase. Other examples of co-ordinate gene expression, of which some have been cited as examples of eukaryoptic operons in the past, are probably more easily explained as examples of close linkage of two gene loci on the same chromosome and their consequent susceptibility to co-ordinate inactivation by deletion or position effect.

Table 7.2 Susceptibility of liver enzymes to hormonal regulation. (From Trueman D. E. S. (1974) *The Biochemistry of Cytodifferentiation*. Blackwell Scientific Publications, Oxford.)

	Glucagon	Hydrocortisone	Thyroxin
Phosphoenolypyruvate carboxylase E.C.4.1.1.32	+		
Tyrosine aminotransferase E.C.2.6.1.5	+		
L-Serine dehydratase E.C.4.2.1.13	+		
Glucose 6-phosphatase E.C.3.1.3.9	+		+
UDPG-glycogen glycosyl transferase E.C.2.4.1.11		+	
Arginase E.C.3.5.3.1		+	+
NADPH-cytochrome c dehydrogenase E.C.1.6.99.2			+
Ornithine aminotransferase E.C.2.6.1.13		+	
Malate dehydrogenase (NADP) E.C.1.1.1.40			+
Tryptophan oxygenase E.C.1.13.1.12		+	
Pyruvate kinase E.C.2.7.1.40			+

So despite the very large amounts of DNA present in eukaryotes as compared to prokaryotes, the individual structural genes are apparently each separately regulated. But although the amount of DNA in eukaryotes is very large, there are strong grounds for believing that the total number of structural genes does not exceed 50 000 and may be as low as 5–10 000. This compares interestingly with the maximum number of possible genes based purely on the amount of DNA—for man the figure is close to a possible 4 million. When the apparent redundancy of eukaryotic DNA is considered alongside the evidence for much nuclear RNA being of high molecular weight, it is plausible to assume that much of the DNA in eukaryotes is regulatory, so that each structural gene would be supported by numerous control genes, each sensitive to different effector molecules and interacting together to ensure that the structural gene is expressed only at the appropriate times. In the absence of any really unequivocal evidence, it is necessary to resort to relatively hypothetical models, of which we can afford space only to a few which seem particularly noteworthy.

7.8.5 Hypotheses of Gene Control in Eukaryotes

We now list a number of interesting models put forward by various laboratories.

Only the names of the individuals chiefly concerned with these proposals are listed.

Many of the models which follow have much in common, and almost all have been strongly influenced by the need to explain the very large genomes found in eukaryotes. It is also interesting to notice how the evidence discussed in Chapter 6 has greatly affected the issue. Since the hypotheses frequently overlap with one another in the details they cannot easily be sorted out into distinct categories. Nor can they usefully be discussed chronologically because the authors have often continued to revise them with the passage of time and the appearance of new evidence. But although new evidence may have rendered a model partially or completely obsolete, even in the mind of its author, a consideration of these theories serves to underline the problems which we are now considering.

COMINGS

Comings has coined the term 'junk' DNA, and according to his proposal eukaryotes are simply saddled with rather large quantities of essentially useless DNA. It is suggested that varying amounts of this material may or may not be transcribed but it is, in any event, neither translated into useful protein, nor involved in useful regulation.

CALLAN

This model has been given the title of the 'master and slaves' hypothesis, and is discussed in Chapter 6. It was derived largely from studies on lampbrush chromosomes. The suggestion is that many or most structural genes are present as multiple copies, one copy being a master copy and the others so-called slave copies. Any mutations in the master gene are passed on to the slaves but mutations in the slaves are rectified by reference to the master copy. The correction mechanism is necessary to account for the 'single-hit' evidence from studies on radiation induced mutations, and is assumed to occur frequently. This hypothesis, attractive in many ways, has been rendered unlikely by the strong evidence from hybridization studies which indicates that most genes occur as only one or two copies per haploid genome. But some such correction mechanism might still apply to the multiple copies of the ribosomal genes.

EDSTROM AND LAMBERT

Also stemming largely from studies on transcriptionally active chromosomes, this hypothesis proposes that a large amount of the genomic DNA in eukaryotes is transcribed only during meiosis, and consists of varying copies of homologous loci which represent stages in the evolution of a particular structural gene. Again, the hybridization data argue against such a model. However, it is interesting to note that evidence from gene sequencing studies in a number of laboratories suggests that 'silent' or 'dead' genes may be a common feature of the eukaryotic genome and may represent outmoded or highly divergent copies of normal structural genes. Edstrom and Lambert's model seems to anticipate some of these data.

It should be noted that none of these three hypotheses actually proposes control mechanisms, but simply tries to account for apparently superfluous DNA. The other models to be considered are more concerned with the details of control.

GEORGIEV

This proposal was one of the originals in the field and still retains interest. It assumes that each structural gene is associated with a long series of regulatory genes to form a transcriptional unit, thus explaining the very large size of much nuclear RNA. The polymerase enzyme would traverse all of these control genes, each of which would have to be individually switched on by positive or negative control, before transcribing the single copy of the structural gene. It follows, according to Georgiev's ideas, that much or most of the DNA in the eukaryotic genome is made up of control sequences, and each structural gene would have many control sequences each sensitive to different regulatory molecules. So instead of many genes being linked together in operons, the control genes are actually duplicated in overlapping sets for each structural gene.

BRITTEN AND DAVIDSON

This hypothesis proposes that eukaryotic gene expression consists of a hierarchical system of control, with RNA as well as protein providing regulatory molecules. They therefore differ from Georgiev, who envisaged all control genes being eventually transcribed into the primary transcriptional unit and therefore functioning by blocking the passage of the polymerase. Britten and Davidson visualize some control genes providing intermediates, either RNA or, on translation, protein, which will act on other regulatory sequences not necessarily adjacent to them, (see Fig. 7.7). They have introduced a useful series of terms to explain their model which are as follows.

Sensor Genes

Sequences which bind agents which induce patterns of gene expression, directly or indirectly. These, then, are the genes which detect hormones and other signalling molecules which settle on or enter the cell.

Integrator Genes

These genes are postulated to produce regulatory RNA (activator RNA in Britten and Davidson's terminology) following the activation of an adjacent sensor gene. The regulatory RNA has the power to initiate a response from many separate genes, thus the term integrator gene.

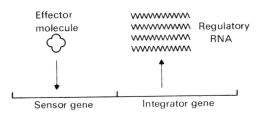

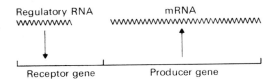

Fig. 7.7 The Britten and Davidson Hypothesis of eukaryotic gene regulation.

Receptor Genes

The sequences which specifically bind the regulatory activator RNA are termed receptor genes, and just as the sensor gene is the receptor site for the integrator gene, so the receptor gene is the site for the structural gene itself. The actual structural genes which produce RNA which is non regulatory, i.e. messenger, transfer, or ribosomal RNA, are styled *producer genes* in Britten and Davidson's model. One of the interesting aspects of this hypothesis (and some other models which are discussed) is that some of the genes, e.g. receptor genes, will exist as numerous scattered identical (or near identical) copies, and would therefore fall into the category of moderately repetitious DNA sequences.

PAUL

The model which Paul has proposed has similarities to Britten and Davidson's. Paul argues that gene transcription is normally accompanied by, or preceded by, unfolding of a supercoiled mass of nucleohistone. He suggests that this unfolding is accomplished by polyanions adhering to special sites on the DNA, termed 'address' loci. When an address locus is so destablized, the adjacent chromatin decondenses and permits transcription of structural genes within it. Paul further suggests that the large amount of DNA in a chromomere or polytene chromosome band is made up of numerous gene copies each equipped with their own address loci. In some cases e.g. the ribosomal genes, these multiple copies will all be utilized and transcribed. More commonly, all but one or two will have deviated by evolution and will no longer provide useful RNA. If transcribed, their RNA products will be destroyed and not utilized as message. It is perhaps this message-selection aspect of the theory which is most difficult to explain, although it is known that much RNA is degraded within the eukaryotic nucleus.

This hypothesis is in some aspects close to that of Paul. It also stresses the connection between decondensation and transcription, but rather than an 'address' locus being one short sequence, this hypothesis assumes that numerous 'address' loci occur and make up much of the chromomeric DNA. It therefore envisages only one or two structural genes per chromomere, with a few operator-like control genes, the rest of the DNA actually functioning only by specific condensation and decondensation. This last type of DNA these authors have termed 'packing DNA' stressing its susceptibility to external effector molecules such as ions, and somewhat resembling Pauls' 'address' loci. The large size of much nuclear RNA is visualized in this model as resulting chiefly from the transcription of packing DNA, of which only the structural gene sequence will be ultimately conserved (see Fig. 7.8).

It is noteworthy that recent results obtained in the laboratory of Groudine and Weintraub at Seattle lend support to the idea that active genes, and also the DNA flanking active genes, are in an extended conformation, whilst genes not normally expressed in a particular cell type are condensed. These authors and their collaborators have exposed chromatin to gentle digestion with the nuclease enzyme DNase I, then assayed with a probe for the presence or absence of particular sequences. Their findings are that cells in which a particular gene is active are readily amenable to digestion of that gene, that is, that the gene is more available for digestion in cells in which it is active. They have also extended their experiments to show that sequences flanking active genes are also prone to digestion. This would seem to give credence to the idea of 'packing DNA' as suggested by Maclean and Hilder. Although the hypothesis of packing DNA was made before the discovery of introns, it may also be that many or most intron sequences function as 'packing DNA'.

Less general than the other hypotheses which we have discussed, this proposal lays great stress on the permanence of much cell differentiation and the significance of the number of mitoses experienced by a cell in determining its fate. These workers suggest that this cellular time base may be due to enzymic modification of specific bases in the DNA. It is supposed that the susceptible bases are located at protein binding sites such as operators and promoters, and that modification alters the affinity of such sites for their binding proteins. Specificity depends on the presence of recognition sequences, near to the relevant bases, which would bind or attract the modifying enzyme. The base modification envisaged would be such reactions as methylation, and once executed, many of the modifications could be passed on from cell to cell through successive generations. Although not an implicit part of their model, we should notice that modifications of DNA (and also of the histones bound to DNA) is a

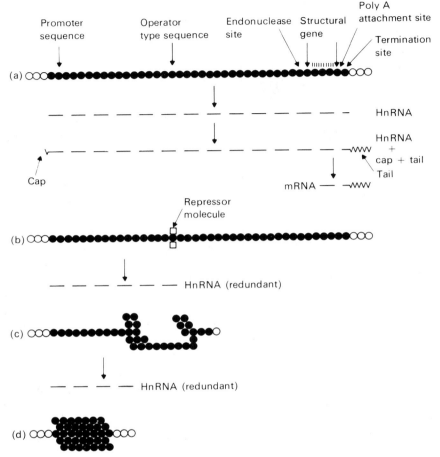

Fig. 7.8 The Maclean and Hilder Hypothesis of eukaryotic gene regulation.

(a) shows an entirely decondensed 'chromomere' or transcriptional domain and the post-transcriptional processing of the HnRNA to give mRNA.

(b) A transcriptional domain in which the RNA polymerase molecule is blocked by a repressor and cannot transcribe the structural gene. Only redundant RNA is produced.

(c) In this figure the packing DNA of the domain in partly condensed, blocking the RNA polymerase after transcription of a very short piece of 'non-coding' DNA.

(d) Here the chromomere or transcriptional domain is completely condensed and no transcription can occur.

potentially powerful way of introducing permanent specificity, and that modification enzymes, with special properties of sequence recognition, are known to occur and operate in bacteria. Evidence is accumulating, in eukaryotes, that DNA sequences that are being transcribed are very low in 5-methyl-cytosine with a corresponding excess of 5-methyl-cytosine in non-transcribed sequences.

7.8.6 Some General Comments about the Hypotheses

The hypotheses which have been mentioned are only a selection from a larger number of postulates which have been advanced. Some are now old (1970) and others fairly new (1979) and all are built on a mixture of established fact and pure speculation. Their quotation here must not be taken to imply that the authors of these theories still sustain belief in them, but nevertheless they do provide a very useful framework within which to discuss gene regulation.

We will briefly review such salient features and evidences of eukaryotic gene organization which, unlike the hypotheses built on them, cannot be seriously questioned. Some of this has been discussed in depth in earlier chapters, especially Chapter 4.

(1)   Eukaryotes possess very large genomes, much larger than they appear to require for strict genetic purposes. Moreover some closely related organisms have highly disparate amounts of DNA (see Table 7.3).

(2)   Much of the primary transcriptional product of the eukaryotic genome is very long, and most of this product is degraded within the nucleus.

(3)   Most eukaryotic structural genes are present as one or very few copies, judging from the hybridization data.

(4)   The DNA of eukaryotes consists of highly repetitious, moderately repetitious and unique sequences. The moderately repetitious sequences are commonly interspersed between the unique sequences and exist largely as 'families' of related or identical sequences (see Table 7.4).

(5)   A DNA tract could be recognized by its base sequence or by a three dimensional structure adopted by the folding of a single chain.

(6)   Some eukaryotic DNA is organized into chromomeres and in these situations localized genetic activity is frequently correlated with decondensation of chromatin.

(7)   Recent evidence from work on the genes for immunoglobulins, globin, ovalbumin and others demonstrates that all these genes are interrupted by quite long insert sequences within the structural gene sequence, yet these inserts are not represented in the relevant mRNA or protein (see Table 7.5). Some eukaryotic genes consist of more than a score of separate pieces and only a few such as histone, have been as yet proven to be uninterrupted. At present most evidence favours the idea that the inserts are transcribed and the resulting HnRNA then trimmed down to size enzymatically. The pieces of a single mRNA molecule are then spliced together within the nucleus by ligase enzymes. It is still premature to attempt to measure the impact of this new evidence on ideas of eukaryotic gene regulation. It may be that some of the inserted sequences—introns—are themselves regulatory, but many appear to be simply redundant. As has been pointed out by P. M. B. Walker, it is probable that the length and sequence of non-coding introns need not be stringently conserved, provided that the sequences which form the bridge when the intron is cut out, are able to make an effective junction. Variation in the size and sequence of homologous introns may help to

Table 7.3 Nuclear DNA amounts in a variety of eukaryotes and prokaryotes. 1 picogram $= 10^{-12}$ g. (From Maclean N. (1977) *The Differentiation of Cells*, Edward Arnold, London.)

Organism	Amount of nuclear DNA		
Bacteriophage I1	0.000007 picograms per particle		
,, T4	0.000025 ,,	,,	,,
E. coli	0.009 ,,	,,	cell
Bacillus megatherium	0.070 ,,	,,	,,
Neurospora crassa	0.017 ,,	,,	,,
Aspergillus nidulans	0.044 ,,	,,	,,
Lupinus albus	2.3 ,,	,,	2C nucleus
Rumex sanguineus	3.1 ,,	,,	,,
Pisum sativum	9.1 ,,	,,	,,
Zea mays	15.4 ,,	,,	,,
Allium roseum	20.4 ,,	,,	,,
Allium karataviense	45.4 ,,	,,	,,
Lilium longiflorum	134 ,,	,,	,,
Drosophia	0.2 ,,	,,	,,
Chiton	1.3 ,,	,,	,,
Nereid worm	2.9 ,,	,,	,,
Squid	9.0 ,,	,,	,,
Gryllus domesticus	12.0 ,,	,,	,,
Protopterus	100.0 ,,	,,	,,
Salmo irideus	4.9 ,,	,,	,,
Esox lucius	1.7 ,,	,,	,,
Tetrodon fluviatilis	1.0 ,,	,,	,,
Amphiuma	168.0 ,,	,,	,,
Rana esculenta	16.8 ,,	,,	,,
Rana temporaria	10.9 ,,	,,	,,
Alligator	5.0 ,,	,,	,,
Black racer snake	2.9 ,,	,,	,,
Pheasant	1.7 ,,	,,	,,
Chicken	2.3 ,,	,,	,,
Ox	6.4 ,,	,,	,,
Man	6.0 ,,	,,	,,
Mouse	5.0 ,,	,,	,,

explain the great differences which are observed in the overall DNA content and composition of quite closely related species.

(8)   Cloning of sequences of eukaryotic DNA has revealed another interesting fact, that the promoter sequence may often be within the structural gene itself. In the laboratory of Donald Brown, DNA flanking sequences on either side of the *Xenopus* 5S-Structural gene sequence have been gradually removed without detectable loss of promoting activity. This probably implies that the

Table 7.4 Details of the Genomes of various organisms showing types of sequence interspersion. (From Davidson E. H. *et al.* (1975) *Chromosoma* **51**, 273.)

Organism	Genome size (pg per haploid genome)	Single copy fraction of DNA	Single copy complexity (nucleotide pairs)	Repetitive sequence classes (average occurrence of each sequence per haploid genome)	Are repetitive sequences interspersed at length given in column 9?	Fraction of repetitive DNA in 200–400 nucleotide elements	Fraction of single copy sequence interspersed with repetitive sequence at length given in column 9	Fragment length (nucleotides)	References
Spisula solidissima	1.2	0.75	8.2×10^8	30	yes	0.60	>0.70	2,300	Goldberg *et al.*, 1975
Crassostrea virginica	0.69	0.60	3.8×10^8	3,700 40	yes	0.35	>0.75	3,000	Goldberg *et al.*, 1975
Aplysia californica	1.8	~0.40	10.7×10^8	85	?	0.60	>0.80	2,500	Angerer, Davidson, and Britten, in prep.
Limulus polyphemus	2.8	0.70	17.9×10^8	4,600 50	yes	0.75	>0.70	2,000	Goldberg *et al.*, 1975
Cerebratulus lacteus	1.4	0.60	7.7×10^8	~2,000 40 1,000	yes yes probably	0.55	>0.70	2,800	Goldberg *et al.*, 1975
Aurelia aurita	0.73	0.70	4.7×10^8	180	yes	0.60	>0.80	2,000	Goldberg *et al.*, 1975
Strongylocentrotus purpuratus	0.89	0.75	6.1×10^8	100 1,500	?	0.75	0.7	3,300	Graham *et al.*, 1974
Xenopus laevis	2.7	0.75	18.5×10^8	100 2,000	yes ?	0.75	0.7	3,700	Davidson *et al.*, 1973, 1974
Rattus norvegicus	3.2	0.75	22.3×10^8	low frequency? 1,800	? yes	?	>0.65	3,200	Bonner *et al.*, 1974; Holmes and Bonner, 1974; Pearson, Wilkes & Bonner, unpublished data
Drosophila melanogaster	0.12	0.75	0.82×10^8	35	no	0.1	non observed	2,500	Manning *et al.*, 1975

Chapter 7

Table 7.5 A Current list of Eukaryotic genes known to possess introns (August, 1979). (Kindly prepared by Dr S. Gregory.) Footnote to table on opposite page.

Gene	Organism	Introns (no. and size)
1. β-globin	rabbit	2 ($\sim$600 and 125 bp)
2. β-globin (maj. or min.)	mouse	2 (646 + 116)
3. α-globin	mouse	2 ($\sim$150 + 100) (positions same as for β-maj.)
4. β-globin (adult)	chicken	2 ($\sim$800 + 100) (same positions as mammalian β-g genes)
5. β-like globin (embryonic)	chicken	1 ($\sim$800)
6. β-globin	human	
7. δ-globin	human	
8. $^G\gamma$-globin $^A\gamma$-globin	human	1 ($\sim$1000)
9. Ovalbumin	chicken	7 (250–1500)
10. 'Pseudo-ovalbumin'	chicken	2(?)
11. 2 ovalbumin-associated genes	chicken	7 each
12. Conalbumin	chicken	16
13. Ovomucoid	chicken	7 (125–1750)
14. Lysozyme	chicken	3 (1250, $\geqslant$900, < 300)
15. Serum albumin	rat	13 (300–1500)
16. Growth hormone	rat	?
17. Immunoglobulin light chain λ_1 and λ_{11}	mouse	2 (1250 in C gene, 93 in V gene)
18. Igκ-light chain (multiple V_κ and J_κ genes, only one C_κ, C_λ, V_λ, J_λ)	mouse	1 in C gene
19. Ig α-heavy chain	mouse	2 (both 100–200)
20. 28S ribosomal (two types)	*Drosophila*	I. 2 (5000 + 500–1000) II. 1 (1500–4000)
21. 25S ribosomal	*Tetrahymena pigmentosa*	1 (400)
	strain 6UM (not 8ALP)	No apparent difference in rRNA or growth of strains with or without insert.
22. 24S ribosomal	*Neurospora* mitochondria	1 (2000–2500)
23. 21S ribosomal	yeast mitochondria	1 (1160)
24. Cytochrome b	yeast mitochondria	4 (600–2000)
25. Cytochrome c oxidase subunit 1	yeast mitochondria	2 or 3
Mitochondrial inserts vary in size in related *Saccharomyces* strains.		
26. tRNA^{Tyr}	yeast	1 (14)
27. tRNA^{Phe}	yeast	1 (18 or 19)
28. tRNA^{Trp}	yeast	1 (34)
29. tRNA^{Tyr}	*Xenopus*	1

RNA polymerase enzyme recognizes the promoter and then stretches or moves upstream to commence gene transcription. This observation does little to strengthen the view that most eukaryotic DNA is regulatory! Brown's laboratory has also isolated another protein which acts as a co-factor in polymerase interaction with this site.

(9)   There is now a lot of circumstantial evidence which indicates that some non-histone proteins play a specific role as gene regulator in eukaryotes. Attempts to isolate such proteins are in progress but results are still preliminary. Experimental systems which permit their isolation, include the use of isolated nuclei as transcriptional assays for cytoplasmic fractions. Insect polytene nuclei are so used by Compton and McCarthy in the United States and *Xenopus* erythrocyte nuclei in the laboratory of one of us (N.M.) at Southampton, England. It should also be feasible to test such proteins, when isolated, by their microinjection into oocytes or eggs.

It does seem sensible to stress that, although regulatory molecules have been isolated and identified in bacteria, and all such have so far proved to be proteins, there is no good reason yet for rejecting RNA as a strong contender for some or even most eukaryotic gene regulation.

One of the neatest pieces of visual evidence supporting much of what has been stressed in the last few paragraphs is contined in a recent publication from Callan's laboratory at St Andrews (Old, Callan & Gross 1977). As illustrated in Figs 7.9 and 7.10, these authors have been able to assemble a radioactive probe for histone gene sequences by making 'hot' histone gene DNA from 'nicked' plasmid DNA which was known to contain histone gene DNA from sea urchins. Details of plasmid technology and insertion of genes in plamids is discussed in Chapter 3. Hybridization of this 'hot' histone DNA with preparations of lampbrush chromosomes from newts results in highly specific labelling of RNA on certain loops.

Unfortunately there is some doubt about whether the labelling is faithful, or whether a tail of non-histone sequence on the probe may account for some of the hybridization. The number of labelled sites in the lampbrush chromosome set appears to be less than ten, but it is not possible to determine whether those sites are single gene loci or contain multiple copies of the same or different genes. The pattern of labelling on any one loop can be followed and, as shown in Fig. 7.11, the histone gene is seen to be associated with a much larger transcriptional

---

No inserts in *Xenopus* tRNAs for Phe, Met$_A$, Met$_B$, Asn, Ala, Leu, Lys.

Values stated in this table are usually approximate (unless the gene has actually been sequenced) and the number of introns discovered is a minimum amount. Reports of introns in nuclear RNA (and hence probably the corresponding gene) have not been recorded. It should be borne in mind that split genes were first discovered in the animal viruses where it seems to be a common feature, e.g. SV40, Ad2, ASV and polyoma. Introns are not present in sea urchin histone genes, *Xenopus* 5S and most tRNA genes or the yeast mitochondria gene for subunit 9 of the ATPase complex. (Data taken from various sources.)

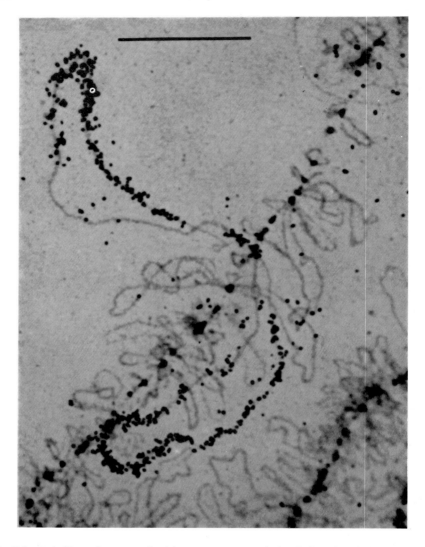

Fig. 7.9 Labelling of presumptive histone gene transcription in lampbrush chromosomes from a newt. A radioactive probing DNA, complementary to histone mRNA, has been utilized for hybridization followed by autoradiography (see however the reservation expressed in text regarding authenticity of labelling). Photograph kindly supplied by Dr R. W. Old. (From Old R. W. *et al.* (1977) *J. Cell Sci.* **27**, 57.)

product which helps to account for the large size of the lampbrush loop. In some loops transcription seems to persist even after the supposedly useful histone message has been entirely transcribed and lost from the end of the remaining tail of recently synthesized RNA.

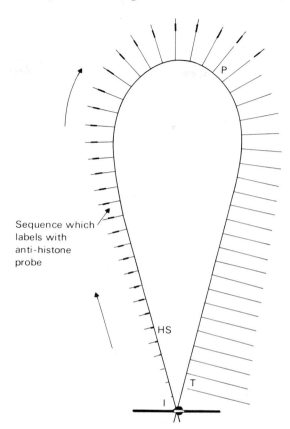

Fig. 7.10 Sequential labelling of lampbrush loop by the technique shown in Fig. 7.9. This diagram indicates what can be visualized on some loops, showing that the histone message comprises only a small part of the complete transcript, and is actually cut off before transcription of the entire loop is completed. Photograph kindly supplied by Dr R. W. Old. (From Old R. W. *et al.* (1977) *J. Cell Sci.* **27**, 57.)

7.9 FURTHER ASPECTS OF GENE REGULATION IN EUKARYOYES

7.9.1 Actions of Hormones in Gene Expression

It is now well established that hormones profoundly affect the pattern of gene expression in their target tissue, although this is normally a quantitative rather than qualitative effect on the activity of individual gene loci. Many hormones, particularly the large polypeptide hormones such as insulin, operate via a 'second messenger system' involving cyclic AMP, and the range of cellular effects now known to be mediated by cyclic AMP is staggering (see Table 7.6). The precise interaction between cyclic AMP and the individual gene sequence is still poorly understood, although it is interesting to recall that an early part of the promoter sequence of the '*lac*' operon in *E. coli* is a cyclic AMP CAP recognition site. Perhaps similar sites exist on the promoters of many eukaryotic genes, so affecting their effective transcription by the RNA polymerase enzyme.

Table 7.6 Hormone actions mediated by changes in levels of cyclic AMP. (In Litwack G. (ed.) (1972) *Biochemical Actions of Hormones Vol. II.* Academic Press, London.)

Hormone	Tissue	Effect
	Increased cyclic AMP levels	
Adrenocorticotropic hormone	Adrenal cortex	Steroidogenesis
	Fat (rat)	Lipolysis
Luteinizing hormone	Corpus luteum, ovary, testis	Steroidogenesis
	Fat	Lipolysis
Catecholamines	Fat	Lipolysis
	Liver	Glycogenolysis, gluconeogenesis
	Skeletal muscle	Glycogenolysis
	Heart	Inotropic effect
	Salivary gland	Amylase secretion
	Uterus	Relaxation
Glucagon	Liver	Glycogenolysis, gluconeogenesis, induction of enzymes
	Fat	Lipolysis
	Pancreatic β-cells	Insulin release
	Heart	Inotropic effect
Thyroid stimulating hormone	Thyroid	Thyroid hormone release
	Fat	Lipolysis
Melanocyte stimulating hormone	Dorsal frog skin	Darkening
Parathyroid hormone	Kidney	Phosphaturea
	Bone	Ca^{2+} resorption
Vasopressin	Toad bladder, renal medulla	Permeability
Hypothalamic releasing factors	Adenohypophysis	Release of trophic hormones
Prostaglandins	Platelets	Aggregation
	Thyroid	Thyroid hormone release
	Adenohypophysis	Release of trophic hormones
	Decreased cyclic AMP levels	
Insulin	Fat	Lipolysis
	Liver	Glycogenolysis, gluconeogenesis
Prostaglandins	Fat	Lipolysis
	Toad bladder	Permeability
Catecholamines (α-adrenergic stimuli)	Frog-skin	Darkening
	Pancreas	Insulin release
	Platelets	Aggregation
Melatonin	Frog skin	Darkening

Certainly some small metabolites such as AMP and GTP interact specifically with DNA either directly or via an interaction with a specific protein.

Steroid hormones, in contrast to polypeptide hormones, pass through the cell membrane and, in many cases, are known to enter the nucleus. It is widely supposed that they interact with the chromatin and may well operate as specific gene regulatory molecules. The Britten and Davidson model of eukaryotic regulation would envisage such hormones interacting with sensor gene sequences. Certainly, in some experimental situations, steroid hormones seem to initiate fairly specific changes at a transcriptional level. Thus, the effects of oestrogen and progesterone on chick oviduct have been found to involve the production of novel RNA in the nucleus, and, as discussed in Chapter 6, the insect steroid hormone ecdysone has specific effects on the puffing of individual gene loci. But until more is known about eukaryotic chromatin, the '*modus operandi*' of steroid hormones on gene activity and expression must remain conjectural. O'Malley's laboratory in Texas is currently attempting to clone the steroid hormone receptor gene which regulates the chick ovalbumin gene.

7.9.2 Modification of Chromatin and DNA

It is often assumed that the structure of DNA is so stable that alterations in the sequence are exceedingly rare. Such alterations in sequence include, of course, mutations, and these form the basis for evolutionary change through natural selection. But we should pause to ask whether mutation is indeed always such a rare event that its occurrence in somatic cells can be ignored in terms of day to day regulation. It is safe to assume that the genomes of all differentiated cells within one organism remain identical and unaltered? All cells are provided with a highly efficient DNA repair service, and any mutation which becomes effective is, by definition, a change in the DNA sequence which has evaded repair. The repair process commonly involves two kinds of enzymes, excision enzymes which cut out a length of DNA which contains a mismatching base or bases, and repair polymerases which insert new suitable bases. It should also be noted that various strains of bacteria are found to have very different tolerances to ionizing radiation (which induce mutations), different repair efficiences and therefore different mutation rates. Indeed differences in susceptibility to mutation may vary from one gene locus to another in the same organism (although it should be stressed here that it is easy to confuse detectable mutation, which is normally an altered enzymic activity or morphological character, with true mutation. Most mutations will remain undetectable to the experimentalist since they will not alter protein structure at an active site.) Differences in DNA repair efficiency are also apparent in eukaryotes and some interesting examples are recognized in man. Thus, persons suffering from Xeroderma pigmentosa, a disease involving excessive sensitivity of the skin to sunlight and an increased incidence of cancer of the skin, are known to have defective excision repair enzymes. Similarly progeria, a rare disease which presents as premature aging, seems to involve deficient repair of single strand breaks in DNA, while ataxia telangiectasia, another rare and inherited disease in which affected individuals develop tumours following X-irradiation, also seems to involve defective repair of the radiation damage to DNA.

Now all of this must make us aware that mutation rate itself is under considerable selective pressure and can be increased or decreased by affecting the efficiency of repair. And if the incidence of actual DNA damage is much commoner than the observed incidence of mutation, could it be that somatic mutation is utilized as a device in differentiation to either silence the activity of certain gene sequences, or to alter their activity by permitting mutation of their regulatory genes? It is certainly possible.

Besides actual mutation of the gene itself, another mechanism which has been proposed to help explain altered patterns of gene activity is modification of particular DNA bases by methylation of cytosine. This is an important part of the gene regulation theory of Holliday and Pugh. Such changes would be controlled by modification enzymes and could be reversible. Some would even result in true mutation. It has been further proposed that histones responsible for stabilizing particular gene sequences might be modified by somewhat similar processes, thus altering the activity of the gene with which they are associated. Such modifications would go some way towards overcoming the basically conservative structure of histones and render them more flexible in terms of specific interactions with DNA in chromatin (see discussion in Chapters 1 and 2).

7.9.3 Multiple Gene Copies

Some genes are present in eukaryotic genomes as multiple copies, presumably in order to permit massive output of the RNA produce. The best known examples are the genes coding for transfer and ribosomal RNAs. In man, for example, there are about 1300 copies of the transfer RNA genes (of which there are, of course, more than twenty different sequences), 2000 copies of the gene for 5s ribosomal, and 280 of the 18s and 28s ribosomal gene see Table 7.7. There are also two interesting phenomena which affect the control of expression of eukaryotic ribosomal RNA genes and which we should therefore mention here. The first is termed *gene rectification*. This process is best known in the fruit fly *Drosophila*, in which certain mutants were found to have lost a major block of the ribosomal RNA gene complex. However, progeny from crosses of affected flies contained a high proportion of individuals with a spontaneous reversion to the normal wild type ribosomal RNA gene number, even when both parents were homozygous for the deletion. This was termed gene magnification. A return to the wild type dose of rRNA genes is also displayed by the progeny of flies with increased rRNA genes following from genetic duplication of the nucleolar region. So gene rectification may involve reduction as well as increase. We can

Table 7.7 Numbers of genes coding for transfer and ribosomal RNAs in different organisms. (From Gurdon J. B. (1974) *The Control of Gene Expression in Animal Development.* Clarendon Press, Oxford.)

Species	tRNA	No. of genes coding for 18S/28S rRNA	5S rRNA	Size of genome	Organism
Bacterium	50	6	7–14	2.8×10^9 daltons	*E. coli*
Bacterium	42	9–10	4–5	3.9×10^9 daltons	*B. subtilis*
Yeast	320–400	140	—	1.25×10^{10} daltons	*S. cerevisiae*
Insect	860	180–190	195–230	1.2×10^{11} daltons	*D. melanogaster*
Frog	1150	450	24 500	1.8×10^{12} daltons	*X. laevis*
Mammal	1310	280	2000	2.1×10^{12} daltons	Hela (human)

surmise that while the activity of other gene sequences is controlled quantitively by cyclic AMP and various regulatory molecules, the ribosomal genes are required to be fully active much of the time. Thus, they are present in many copies per genome and the correct number seems to be itself rather carefully regulated.

The second phenomenon *gene amplification*, is one of the most exciting discoveries of molecular biology. It has been estensively studied in *Xenopus laevis*, but is also known in many insects, molluscs, fish and amphibians. Gene amplification is a temporary but dramatic increase in the numbers of gene copies for the major rRNAs, and is found in the developing oocyte. During oocyte maturation in these species a greatly increased number of nucleoli is present—sometimes many thousands, most being free from the chromosomes and each containing a core of the DNA block of 18s and 28s gene copies. The phenomenon was first described by Gall in 1968. Whereas the number of gene copies of the major rRNA genes in somatic tissues of *Xenopus laevis* is about 2000, in the oocyte this rises to about 2×10^6, involving a thousand-fold increase. It must be emphasized that the phenomenon is normally transient, being limited to the egg and early embryo, although more permanent multiplication of the ribosomal genes is observed in some species such as *Drosophila*—thus the need for gene rectification. The only other known example of natural gene amplification is of the genes for chorion proteins in egg follicle cells in *Drosophila*.

We should understand gene amplification as a somewhat crude way of regulating the output of the ribosomal RNA genes, and of course the early embryo does have a very great demand for active ribosomes in the first few rounds of cell division. It should be pointed out that some other genes are in a sense, amplified, for example those on the mitochondrial genome, but to avoid confusion the term is best not extended to cover such examples.

Aside from the transfer and ribosomal RNA genes, are any other structural gene sequences present as multiple copies? The answer is that a few are known to be, but the number of copies is invariably low and the different copies are often slightly divergent. Thus, in man there appear to be, in each haploid genome, single copies of the genes for beta and delta globin, and probably two copies each of the alpha and gamma globins, but in each of the latter cases there is slight variation between copies. But the multiple copies of genes for histone in such organisms as sea urchins is no doubt another example of a true multiple copy gene. We can therefore conclude that, as a generalization, structural genes are present as single, or very few, copies per haploid genome. This implies, therefore, that control over the rate of RNA production by these genes must all be engineered by means other than modulating the numbers of gene copies active at any one time. Only in the case of the ribosomal genes is such a regulatory device known to be utilized. Perhaps this is because the possession of more than a very few copies of structural gene sequences would render such genes extremely resistant to mutational change, a situation perhaps best avoided by living organisms in a changing world.

7.9.4 Allelic Exclusion

The genetic basis of the form and function of antibodies (or immunoglobulins, as they are more properly termed) is a highly complex but intriguing subject. Although the greater part of the topic is outside the scope of this book, there is one aspect which certainly deserves mention, namely the phenomenon of allelic exclusion. What this phrase refers to is that, within any single antibody-producing cell, only one of the two allelic forms of the immunoglobulin are synthesized, assuming the organism to be heterozygous for that locus. Moreover, this unilateral expression of the chromosomes is handed down to the progeny of a cell so committed, thus inducing the production of a clone of cells devoted to the production of a single antibody species. To date, no other authenticated example of allelic exclusion in eukaryotes is known. And of course there are many situations where it obviously does not occur. For example, persons heterozygous for the sickle form of the beta chain of adult haemoglobin would, with allelic exclusion, possess some blood cells which sickled easily (where the S allele was active) and some which sickled not at all (where the normal beta allele was active). This is not the case. Persons heterozygous for this gene possess blood cells all of which will sickle slightly on exposure to low oxygen tension.

There are, however, many other situations in which allelic exclusion might apply but has never been adequately sought. So far, only the immunoglobulin system definitely registers.

7.9.5 Episomes and Plasmids in Gene Expression

The form and function of bacterial plasmids has already been discussed in Chapter 3, but there are a few examples of transposable genetic elements in eukaryotic cells which deserve comment. Some of the best known and most actively researched have been discovered in plants of maize and *Antirrhinum*.

In both these plants, certain genes which affect the colour pattern of seed and flower respectively are found to mutate with a very high frequency—many thousands of times the expected background mutation rate. Also, in different strains of these plants, different gene loci are so affected. The most plausible explanation for these curious findings is that there is a transposable genetic element in operation which can be inserted at different positions on the genome and, when inserted, drastically affects the mutation rate of neighbouring loci. Dr Barbara McClintock, who has pioneered the work on maize, has styled these elements 'transposable control elements' and there is certainly no evidence that they code for protein.

THE ROLE OF INSERTION ELEMENTS (IS)
As discussed in Chapter 3, small DNA elements have been discovered in bacteria which are responsible for affecting mutations at certain loci and for gene arrangements in some plasmids. IS elements seem to be able to induce chromosomal rearrangements which lead to new combinations of both structural and regulatory genes in bacteria. At the moment all we can do is to underline the similarity between such bacterial *IS elements* or translocons, and the transposable genetic elements detected in such systems as maize, yeast mating type, immunoglobulin genetics, trypanosome surface antigens, and *Drosophila* insertional mutagens (see McKay (1980) and Flavell (1981) [Chapter 3]).

7.9.6 Post Transcriptional Control

The emphasis of this book is on chromatin and the main burden of this chapter is about gene expression at the primary level of transcription. But the ultimate expression of the genetic message is in the production of specific proteins and the resultant cells and organism which follow. We will not attempt to follow gene expression at all seriously beyond the production of RNA, but it does seem appropriate to hint at some of the mechanisms which operate close to the level of the gene.

It is conceivable, and to some people highly likely, that most of the control of gene expression is actually post transcriptional, as we broadly hinted when discussing transcription on the lampbrush chromosome in Chapter 6. There is some evidence which suggests that globin genes are transcribed, at least to a limited extent, in tissues such as brain or skin which are never sites of significant haemoglobin synthesis. Moreover the complexity of HnRNA does not seem to alter significantly during development, as we might expect if differentiation implied the switching on of previously silent genes.

Now it may be that these 'chinks in the armour' of the conventional view of selective gene expression are simply indications that gene control is leaky, and that genes are not so much absolutely off or on, but rather transcribed either rarely or frequently. When a gene is said to be 'on' it may simply imply that it is being rapidly transcribed. But we must not exclude an alternative explanation, unlikely though it may seem. It is that most genes are transcribed all the time, and selective expression of particular genes in a differentiated tissue means selective survival of relevant Hn and mRNA. Our present ignorance of HnRNA is so great that we cannot presently exclude this possibility, and if the inserted non-message sequences in globin and ovalbumin are found to be normally transcribed, it would strongly suggest that post transcriptional control of gene expression is much more elaborate and significant than was previously imagined.

Post-transcriptional modification of RNA commences immediately or soon after the polymerization of RNA (see Table 7.8). Part of this modification is addition, as occurs in the formation of a tail of poly A sequences on the 3' end of much of the newly synthesized RNA. Some is also capped by the addition of a methylated guanosine residue at the 5^1 end. And as we stated above, there is

Table 7.8 Examples of post-transcriptional pre-translational processing of RNA.

Process	Description
Capping	This involves the attachment of a guanosine residue to 5' end of some nuclear RNA and the subsequent methylation of that G nucleotide. Found on some Hn and mRNA.
Tailing	This consists of the addition of a tail of some 10–20 adenosine residues to the 3' end of some nuclear RNA. Found on some Hn and mRNA and perhaps acts as a ticketing system, the string being shortened as message is recycled.
Base modification	Some RNAs, particularly transfer RNAs, have certain bases modified by methylation after transcription.
Cleavage	This is the enzymatic cutting of RNA in special predetermined sites. It may occur in three different ways: cutting of a long polycistronic message to yield individual messages. e.g. the RNA for vitellogenin; trimming of so called spacer regions at either end to yield functional RNA. This certainly takes place with 28 and 18S ribosomal RNA precursor; cutting or excision within a structural gene sequence to yield a functional message. Occurs in precursor RNA for globin, ovalbumin etc. where introns are present.
Ligation	RNAs such as globin mRNA precursor require introns to be cleaved out, followed by ligation of the pieces, to yield functional mRNA.
Degradation	Much nuclear RNA never enters the cytoplasm and is presumed to be enzymatically degraded as completely redundant.
Ribonucleoprotein aggregation	There is some evidence that mRNA is associated with specific protein before it is used for translation. This may occur within the nucleus or in the cytoplasm.

evidence which indicates that both capping and tailing survive later RNA modification and persist on the mRNA. But cleavage and partial degradation is also a frequent post-transcriptional modification to RNA, some polycistronic RNA being cleaved into separate sections within the nucleus, and indeed very considerable sections of RNA are enzymatically degraded. So, much of the RNA 'turns over' rather rapidly within the nucleus and never functions as a template for translation.

Even if a particular messenger RNA does find its way to the cytoplasm, it may not be readily translated. Some indeed seems to be stored for long periods, often in combination with protein, and even if not stored, ribosome availability and translational efficiency of transfer RNA must clearly affect the speed with which a protein product actually materializes.

7.10 FURTHER READING

Bostock C. J. & Sumner A. T. (1978) *The Eukaryotic Chromosome.* North Holland, Amsterdam.

Brown D. D. (1981) Gene expression in Eukaryotes. *Science* **211**, 667–74.

Burdon R. H. (1974) *RNA Biosynthesis.* Chapman and Hall, London.

Callan H. G. (1967) The organisation of genetic units in chromosomes. *J. Cell Sci.* **2**, 1–7.

Callan H. G. & Old R. W. (1980) *In situ* hybridization to lampbrush chromosomes: a potential source of error exposed. *J. Cell Sci.* **41**, 115–23.

Chambon P. (1978) *The Molecular Biology of the Eukaryotic Genome is Coming of Age.* Cold Spring Harbour Symposium XLII Part 2, 1209–34. (Many other papers in this volume are also of particular relevance to this chapter.)

Davidson E. H. & Britten R. J. (1973) Organization, transcription and regulation in the animal genome: *Quart. Rev. Biol.* **48**, 565–613.

Davidson E. H., Galau G. A., Angerer R. C. & Britten R. J. (1975) Comparative aspects of DNA organization in metazoa. *Chromosoma.* **51**, 253–9.

David I. B. & Wahli W. (1979) Application of recombinant DNA technology to questions of developmental biology: a review. *Dev. Biol.* **69**, 305–28.

Edstrom J. E. & Lambert B. (1975) Gene and information diversity of eukaryotes. *Prog. Biophys. Mol. Biol.* **30**, 57–82.

Georgiev G. P. (1972) The structure of transcriptional units in eukaryotic cells. *Curr. Top. Develop. Biol.* **7**, 1–60.

Holliday R. & Pugh J. E. (1975) DNA modification mechanisms and gene activity during development. *Science* **187**, 226–32.

Humphries S., Windass J. & Williamson, R. (1976) Mouse globin gene expression in erythroid and non-erythroid tissues. *Cell* **7**, 267–77.

Lewin B. (1974) *Gene Expression Vol. 2.—Eukaryotic Genomes.* John Wiley & Sons, London.

Lewis E. B. (1964) Genetic control and regulation of developmental pathways. In *The Role of Chromosomes in Development*, ed. Locke M., p. 231. Academic Press, New York.

Lippincott J. A. (1977) Molecular basis of plant tumour induction. *Nature* **269**, 465–6.

Long E. O. & Dawid I. B. (1980) Repeated genes in eukaryotes. *Ann. Rev. Biochem.* **49**, 727–54.

Maclean N. (1976) *The Control of Gene Expression.* Academic Press, New York.

Maclean N. & Hilder V. A. (1977) Mechanisms of chromatin activation and repression. *Int. Rev. Cytol.* **48**, 1–54.

MacGillivray A. j., Paul J. & Threlfall, G. (1972) Transcriptional regulation in eukaryotic cells. *Adv. Cancer Res.* **15**, 93–162.

McKay P. (1980) Movable genes. *Nature* **287**, 188–9.

Old R. W., Callan H. G. & Gross K. W. (1977) Localization of histone gene transcripts in newt lampbrush chromosomes by *in situ* hybtidization. *J. Cell Sci.* **27**, 57–79.

O'Malley B. W., Towle H. C. & Schwartz R. J. (1977) Regulation of gene expression in eukaryotes. *Ann. Rev. Genet.* **11**, 239–75.

Paul J. (1972) General theory of chromosome structure and gene activation in eukaryotes. *Nature* **238**, 444–6.

Van der Ploeg L. H. T. & Flavell R. A. (1980) DNA methylation in the human globin gene locus in erythroid and nonerythroid tissues. *Cell* **19**, 947–58.

Walker P. M. B. (1979) Genes and non-coding DNA sequences. In *Human chromosomes* pp. 25–45. Ciba Foundation Symposium No. 66 (New Series) Amsterdam, Excerpta Medica.

Wood W. G. *et al.* (1978) Human globin gene expression. *Cell* **15**, 437–46.

Glossary

Active site The region of a protein molecule at which direct interaction with a substrate or regulatory molecule takes place.

Allele (also allelomorph) One member of the range of gene sequences which can occur at a particular locus. Different alleles of the same gene often differ by only one or a few bases.

Allostery A phenomenon displayed by many proteins with two or more receptor sites, and in which occupation of one site alters the specificity of the other available sites.

Angstrom unit (Å) 1×10^{-10} metres, now normally expressed as 0.1 nanometres.

Antibody A protein which neutralizes a specific antigen in a living organism.

Anticodon A group of three nucleotides in a transfer RNA which recognizes three nucleotides of messenger RNA during protein synthesis.

Antigen A molecule that is capable of stimulating the production of neutralizing antibody proteins when injected into a vertebrate.

ATP Adenosine triphosphate.

Autoradiography A technique for the detection of radioactively labelled molecules by overlaying the specimen with photographic film. When the film is developed an image is produced which corresponds to the location of the radioactivity.

Autosome A chromosome other than a sex chromosome.

Bacteriocin Many bacterial strains liberate toxic proteins called bacteriocins which are active only against closely related strains. Toxins of this type produced by *E. Coli* are called *colicins*. Several different types have been isolated which kill sensitive cells by different mechanisms. *E. coli* cells are generally immune to the *colicins* which they produce.

Bacteriophage See 'phage'.

Barr body The densely staining inactivated X chromosome found near to the nuclear membrane throughout interphase in the somatic cells of XX females.

Bivalent A pair of synapsed homologous chromosomes.

cDNA An abbreviation for complementary DNA which is synthesized by reverse transcription from messenger RNA. It is therefore similar to the DNA of the genes but will not possess intron sequences.

C value The amount of DNA in the haploid genome of a eukaryotic cell.

Capped 5'-ends The 5'-ends of eukaryotic mRNAs are modified posttranscriptionally to form ends with the general structure

$$m^7G(5')ppp(5')Nmp. \ldots$$

where m^7G represents a 7-methyl-guanosine residue, and Nm a 2'-o-methylated nucleoside.

Capsid The external protein shell or coat of a virus particle.

Centromere A specialized area of a chromosome which includes the kinetochore, the site of attachment of a chromosome to a spindle fibre.

Chiasma (pl. chiasmata) The crossover between two chromatids, often visible as a connection between two chromatids during prophase I of meiosis.

Chimaera An organism which contains two lines of cells of differing genetic constitution.

Chromatid One of two identical halves of a chromosome, resulting from replication during interphase. Both sister chromatids share a common centromere.

Chromomere Stainable thickened areas arranged linearly along a chromosome.

Cistron A sequence of DNA which specifies one polypeptide chain in protein synthesis.

Clone A group of organisms or cells, all originating from a single ancestral organism or cell, and all genetically identical.

Codon A set of three nucleotides specific for a particular amino acid in protein synthesis.

Col factors Plasmids in bacteria which confer on bacteria the ability to produce colicins.

Constitutive enzyme An enzyme produced continuously.

Cot value A measure of the rate of hybridization or annealling of single stranded DNA. 'Co' refers to DNA concentration and 't' to the time of hybridization in seconds.

Crossing-over The process, visualized by the appearance of chiasmasta, in which reciprocal exchange of genes occurs between two non-sister chromatids of two homologous chromosomes. Non-reciprocal exchange sometimes occurs, and is termed 'unequal crossing-over'.

C-terminus The free alpha carboxyl group of the last amino acid at one end of a polypeptide chain.

Dalton A unit equal to the mass of the hydrogen atom (1.67×10^{-24} g), and therefore an expression of molecular weight.

Diploid A cell or individual with two complete sets of chromosomes.

DNA polymerase Any one of a number of enzymes which catalyze the formation of DNA from deoxyribonucleotides.

Enzyme A substance, normally wholly or mainly protein, which catalyzes a biochemical reaction or regulates the rate of the reaction.

Euchromatin All chromatin other than heterochromatin. Often used simply to refer to non-condensed chromatin.

Eukaryote An organism or cell with a discrete nucleus, in contradistinction with 'prokaryotes' which lack discrete nuclei.

F factor A plasmid or episome in the bacterium *E. coli*, the possession of which confers 'maleness' on the cell.

F1 The first filial generation, that is the first generation resulting from a given cross.

Gene A factor which determines an hereditary trait, now known to be a segment, or part of a segment, of a DNA molecule.

Genome The complete set of genes or chromosomes constituting the genetic endowment of an individual.

Hairpin loop A region of double helix formed by base-pairing within a single strand of DNA or RNA which has folded back on itself.

Haploid A cell or individual with only one set of all the chromosomes. For example unfertilized eggs and sperms are haploid.

Hela cells Cells derived from a carcinoma of the cervix from a woman patient called Helen Lane in 1952. A widely used human tissue cell line.

Heterochromatin Chromatin that is more or less permanently condensed, as in the Barr body of the human female. Sometimes used for any condensed chromatin, as in chromomeres or even condensed portions of interphase chromatin.

Heteroduplex A DNA molecule formed by base-pairing between two strands that do not have completely complementary nucleotide sequences.

Heterozygote An individual with unlike genes in one particular allelic pair. A single individual is therefore likely to be heterozygous with respect to one gene pair, but homozygous with respect to many others.

High Mobility Group Protein (HMG) A group of non-histone low molecular weight proteins which probably play a role in the organization of chromatin structure.

Histone A basic and small molecular weight protein which complexes with DNA.

Homozygote An individual with the same genes in one particular allelic pair. See also heterozygote.

Hybridization The artificial annealling by specific base pairing of complementary or partially complementary sequence lengths of DNA, or DNA with RNA.

Inducer An effector molecule which induces a biochemical reaction to occur.

Induced enzyme synthesis Synthesis of an enzyme only in the presence of a specific inducer or effector molecule.

In vitro Literally 'in glass', meaning in the test tube or outside the organism.

In vivo Within the organism.

Karyotype The chromosome complement of an individual.

Kilobase (abbr. Kb) One thousand nucleotides in sequence.

Linkage Non-random assortment of genes at meiosis due to their occurrence on the same chromosome.

Locus The position on a chromosome occupied by a particular allele or gene.

Lysis Dissolution or disintegration of cells.

Lysogenic bacteria Living bacterial cells which harbour temperate phages and which are therefore not lysed by the phase.

Micron, also styled micrometer, it equals 1×10^{-6} m. Written as μ or μm.

Nanometre (nm) 1×10^{-9} m. Also equal to a millimicron (mμ). The nanometre is now the preferred unit of fine measurement, rather than angstrom or micron.

Nick-translation The formation of a DNA probe by specific nicking of a double stranded DNA sequence by use of enduclease enzyme, followed by insertion of new nucleotides by polymerase activity from the nicked point.

N-terminus The amino (-NH$_2$) end of a peptide chain.

Nucleolar organizer The sequence of DNA within the nucleolus which contains the repeated copies of the genes for the major ribosomal RNA's.

Nucleoside Portion of DNA or RNA composed of a ribose or deoxyribose sugar combined with a purine or pyrimidine base.

Nucleosome The histone DNA aggregate which forms the basic subunit of most chromatin.

Nucleotide Portion of DNA or RNA composed of a ribose or deoxyribose sugar, combined with a phosphate group plus the purine or pyrimidine base.

Okazaki fragment A segment of 1000–2000 nucleotides made as a continuous colinear molecule by a DNA polymerase enzyme. The separate fragments are joined by DNA ligaze enzyme to form the completed DNA chain.

Oligonucleotide A linear sequence of up to ten nucleotides.

Operator A site on a DNA molecule at which a specific repressor protein binds, hence regulating the expression of associated genes within the operon.

Operon Two or more contiguous genes subject to co-ordinate regulation by an operator and repressor.

Peptide bond A chemical bond which links amino acid residues together in protein assembly.

Phage (Also bacteriophage) a bacterial virus.

Phenotype The visible manifestation of the genetic makeup of an individual.

Polynucleotide A linear sequence of many nucleotides.

Polypeptide A compound containing amino acid residues joined by peptide bonds. Proteins consists of one or more specific polypeptide chains.

Polyploid An individual with more than two complete sets of chromosomes.

Polytene chromosome A giant chromosome which is multi stranded, being much broader than a normal chromosome and visible throughout interphase. These chromosomes result from repeated DNA replication without cell division and are found in certain tissues of larval flies.

Position effect A genetic effect resulting from a change in the position of a gene on the chromosome. These effects often result from interactions between neighbouring genes.

Prokaryote A cell or organism which lacks a discrete nucleus.

Probe (hybridization) DNA or RNA molecule radiolabelled to a high specific radioactivity, used to detect the presence of a complementary sequence by molecular hybridization.

Promoter Region of a DNA molecule at which RNA polymerase binds and initiates transcription.

Prophage The state of a phage genome in lysogenic bacterium.

R. factor Bacterial plasmids which carry genes for drug resistance.

Restriction endonuclease A group of enzymes, commonly found in bacteria, which break internal bonds of DNA at highly specific points (often at or near to short palindromic sequences).

Reverse transcriptase An RNA dependent DNA polymerase enzyme.

Satellite DNA A confusing term, normally referring to DNA which bands separately from the DNA of the main genome in caesium chloride centrifugation. Mitochondrial, centromeric and plasmid DNA's may all be obtained as satellites following appropriate procedures.

Sedimentation coefficient (S value) The rate of sedimentation of a solute during centrifugation. Often expressed in Svedberg units which is a S value of 1×10^{-13} second.

Svedberg unit (see sedimentation coefficient).

Synapse The pairing of homologous chromosomes at prophase I of meiosis.

Temperate phage A phage which exists in a bacterium but does not lyse the host in normal circumstances.

Transduction Transfer of bacterial genes from one bacterium to another by a phage particle.

Transformation Genetic recombination resulting in the permanent acquisition of new genes, normally following the incorporation of naked DNA into the cell.

Transposon A DNA element which can insert at random into plasmids or the bacterial chromosome independently of the host cell recombination system.

Translocation The movement of a piece of DNA, often a large piece containing a number of genes, to another part of the same or a different chromosome.

Ubiquitin A widely distributed globular acidic protein with a highly conserved amino acid sequence. It shows a marked affinity for histones and may be involved in chromatin structure.

Viroid A simple infective genetic particle found in certain plants and consisting only of a closed but partly double-stranded circle of DNA.

Virulent phage A bacterial virus which destroys the host cell by lysis.

Index